Disclaimer

Book Title: Development and Verification of a Linear-Fit Mixed System Rating Method for Unitary Two-Speed and Variable-Speed Air Conditioners

Book Author: William V. Payne;

Book Abstract: A linear-fit method of rating residential-type air conditioning systems was evaluated based on performance predictions and laboratory testing of one two-speed matched system and two mixed systems (matched two-speed condensing unit, matched indoor coil blower, and two mixed coil blowers). The individual evaporators and the condensing unit were separately tested using water heated/cooled condensing/evaporating units at standard air conditions over a range of evaporator refrigerant saturation temperatures, evaporator superheats, and liquid refrigerant temperatures. Capacity predictions were within ±1.0 % of the tested values for the mixed systems, and the EER predictions were within ±1.5 % of the measured EERs. The methods used for system rating on the two-speed system can also be applied to a variable-speed system.

Citation: NIST TN - 1667

Keywords: air conditioner; cooling capacity; mixed system; rating procedure; SEER; two-speed system; variable-speed system

Technical Note 1667

Development and Verification of a Linear Fit Mixed System Rating Method for Unitary Two-Speed and Variable-Speed Air Conditioners

W. Vance Payne

U.S. DEPARTMENT OF COMMERCE
National Institute of Standards and Technology
Building Environment Division
Building and Fire Research Laboratory
Gaithersburg, Maryland 20899-8631

National Institute of Standards and Technology
Technology Administration
United States Department of Commerce

Technical Note 1667

Development and Verification of a Linear Fit Mixed System Rating Method for Unitary Two-Speed and Variable-Speed Air Conditioners

W. Vance Payne

U.S. DEPARTMENT OF COMMERCE
National Institute of Standards and Technology
Building Environment Division
Building and Fire Research Laboratory
Gaithersburg, Maryland 20899-8631

June 2010

U.S. DEPARTMENT OF COMMERCE
Gary Locke, Secretary

National Institute of Standards and Technology
Patrick D. Gallagher, Director

TABLE OF CONTENTS

List of Tables

List of Figures

Nomenclature

A   EVAP-COND air-side heat transfer coefficient correction factor

$A_\#$-Test   refers to AHRI Standard 210/240 test conditions of 35.0 °C (95.0 °F) outdoor air and 16.7 °C (80 °F) dry-bulb/ 19.4 °C (67 °F) wet-bulb indoor air conditions with #=1 for low speed compressor, low speed indoor fan and #=2 for high speed compressor, rated speed indoor fan

$B_\#$-Test   refers to AHRI Standard 210/240 steady-state test conditions of 27.8 °C (82 °F) outdoor air and 16.7 °C (80 °F) dry-bulb/ 19.4 °C (67 °F) wet-bulb indoor air conditions with #=1 for low speed compressor, low speed indoor fan and #=2 for high speed compressor, rated speed indoor fan

C_D cyclic degradation coefficient as defined in AHRI Standard 210/240-2003

CD Unit   condensing unit, the outdoor section of the split air-conditioner

CLF   Cooling Load Factor as defined in AHRI Standard 210/240-2003

Diff   abbreviation for difference

DOF   degrees of freedom

EER   Energy Efficiency Ratio as calculated in AHRI Standard 210/240-2003, (Btu/W·h)

ICM   Indoor (independent) coil manufacturer

matched   refers to a split air-conditioning system, an indoor section/condensing unit combination, which rated performance is determined by laboratory testing; also may refer to the evaporator which is used in the matched system.

mixed   refers to a split air-conditioning system, an indoor section/condensing unit combination, which rated performance is not determined by laboratory testing; also may refer to the evaporator which is used in the mixed system.

n   number of tests or number of data points

P electrical power, W

$p_\#(82)$ condensing unit power at $B_\#$-Test condition (indoor fan power not included), W

$P_\#(82)$ total power of air conditioner at $B_\#$-Test condition (condensing unit power plus indoor fan power), W

ΔP EVAP-COND refrigerant-side pressure drop correction factor

Q Cooling capacity, W (Btu/h)

$q_\#(82)$ cooling capacity at $B_\#$-Test condition without accounting for indoor fan heat input, W (Btu/h)

$Q_\#(82)$ cooling capacity at $B_\#$-Test conditions with the indoor fan heat input accounted for , W (Btu/h)

$q_\#(95)$ cooling capacity at $A_\#$-Test conditions without accounting for indoor fan heat input, W (Btu/h)

$Q_\#(95)$ cooling capacity at $A_\#$-Test conditions with the indoor fan heat input accounted for, W (Btu/h)

R   EVAP-COND refrigerant-side heat transfer coefficient correction factor

scfm   standard cubic feet per minute, equivalent to the volumetric flow rate of air with a density of 0.075 lb/ft^3

SEER   Seasonal Energy Efficiency Ratio as defined in AHRI Standard 210/240-2008, Btu/(W·h)

SHR   sensible heat ratio; the ratio of sensible capacity to total capacity

T   temperature

$\hat{\sigma}$ data standard deviation or fit standard error

SSE   sum of squares of the error

ton   cooling or heating capacity equal to 12 000 Btu/h or 3.517 kW

Subscripts

CD   condensing unit of the split system air conditioner
coil   refers to the indoor heat exchanger
cyc   cyclic testing
diff   difference
dry   dry-coil testing
evap   refers to the indoor coil or evaporator at saturated refrigerant conditions
fan   refers to the indoor coil blower
ID   indoor
liq   liquid refrigerant
mixed   refers to the evaporator coil alone with respect to a system
OD   outdoor
ref   refrigerant
ss   steady-state
suph   superheat

NOTE

Use of Non-SI Units in a NIST Publication: The policy of the National Institute of Standards and Technology is to use the International System of Units (metric units) in all of its publications. However, in North America in the heating, ventilation and air-conditioning industry and in the U.S. Department of Energy Test procedure referenced by this document, non-SI units are used instead of SI units; therefore, it is more practical and less confusing to include values in customary units only.

NIST does not approve, recommend, or endorse any product or proprietary material. No reference shall be made to NIST or to reports or results furnished by NIST in any advertising or sales promotion which would indicate or imply that NIST approves, recommends, or endorses any product or proprietary material, or which has as its purpose an intent to cause directly or indirectly the advertised product to be used or purchased because of NIST test reports or results.

Development and Verification of a Linear Fit Mixed System Rating Method for Unitary Two-Speed and Variable-Speed Air Conditioners

W. Vance Payne
National Institute of Standards and Technology

Abstract

A linear fit method of rating residential-type air conditioning systems was evaluated based on performance predictions and laboratory testing of one two-speed matched system and two mixed systems (matched two-speed condensing unit, matched indoor coil blower, and two mixed coil blowers). The individual evaporators and the condensing unit were seperately tested using water heated/cooled condensing/evaporating units over a range of evaporator refrigerant saturation temperatures, evaporator superheats, and liquid refrigerant temperatures. Capacity predictions were within ±1.0 % of the tested values for the mixed systems, and the EER predictions were within ±1.5 % of the measured EERs. The methods used for system rating on the two-speed system can also be applied to a variable-speed system.

Keywords: air conditioner, cooling capacity, mixed system, rating procedure, SEER, two-speed system, variable-speed system

1

ACKNOWLEDGEMENT

This study was sponsored by the United States Department of Energy, Building Technologies Program according to interagency agreement DE-EE0001097 under project managers Wes Anderson and Michael Raymond. Mr. David Wilmering, Mr. Thomas Montgomery, and Mr. John Wamsley provided assistance with test setup, operating the chambers, and analyzing data.

1: INTRODUCTION

A given condensing unit (outdoor section consisting of a condenser, compressor, and associated tubing) is typically offered on the market in several air conditioner models, which differ by the indoor sections they employ. For all models, the manufacturers must provide performance information, which consists of the Seasonal Energy Efficiency Ratio (SEER) and capacity at the 95 °F rating point, Q(95). Federal regulations require that only the highest sales volume indoor-section/outdoor-section combination, referred to as the matched system, be tested in a laboratory to obtain the ratings (CFR 2009a). For other combinations of indoor and outdoor sections, so called mixed systems, the federal regulations allow the use of simplified analytical methodologies upon approval by the U.S. Department of Energy (CFR 2009b).

The most commonly used simplified methodologies for rating mixed systems are those based upon publicly available Q(95) and SEER of the matched systems (e.g., Domanski 1989). The application of these methods requires predicting the capacity of the matched evaporator, which is a major shortcoming because the rater is often not familiar with the matched system product line. Since an inaccurate prediction of the matched evaporator performance leads directly to inaccurate mixed system ratings, a different rating method that excludes this step has the inherent potential to be a better rating approach than the one currently used. Thus NIST developed a single-speed cooling mode linear fit rating method to allow the prediction of the SEER and Q(95) for mixed systems (Payne and Domanski 2006). This method was shown to be able to predict SEER and Q(95) for the tested mixed systems within ±5 % (Payne and Domanski 2005).

Figure 1.1 shows the application of the single-speed linear fit method in a graphical form. This method uses linear fits to the cooling capacity for the mixed coil, and cooling capacities, q(82) and q(95), and power, p(82), for the condensing unit (CD unit). The lines are presented as a function of the evaporator exit saturation temperature. Overlapping of the evaporator and CD unit capacities provides mixed system capacities at 82 °F and 95 °F ambient temperatures. Projecting the saturation temperature corresponding with operation at the 82 °F ambient temperature on the CD unit power chart provides the power requirement for the CD unit at the 82 °F rating point.

In practice the procedure illustrated in Figure 1.1 is performed mathematically. Power and capacity linear fits for the outdoor section are determined by the OEM who also provides EER(95) for their matched system. The ICM then generates linear fits for cooling capacity as a function of evaporator exit refrigerant saturation temperature, T_{evap}, to overlay on the OEM provided outdoor section linear fits. Using the matched system EER(95) and the ICM calculated mixed system EER(95) as shown in Equation 1.1, a mixed system SEER can be calculated using Equation 1.2, where F_{exp} is an expansion device correction factor.

$$EER(95) = \frac{Q(95)}{P(95)} \qquad \qquad 1.1$$

$$SEER_{mixed} = SEER_{matched} \frac{EER(95)_{mixed}}{EER(95)_{matched}} F_{exp} \qquad \qquad 1.2$$

The use of EER(95), instead of EER(82), to determine mixed system SEER is a simplification necessitated by the lack of direct knowledge of the matched system's cyclic degradation coefficient, C_D.

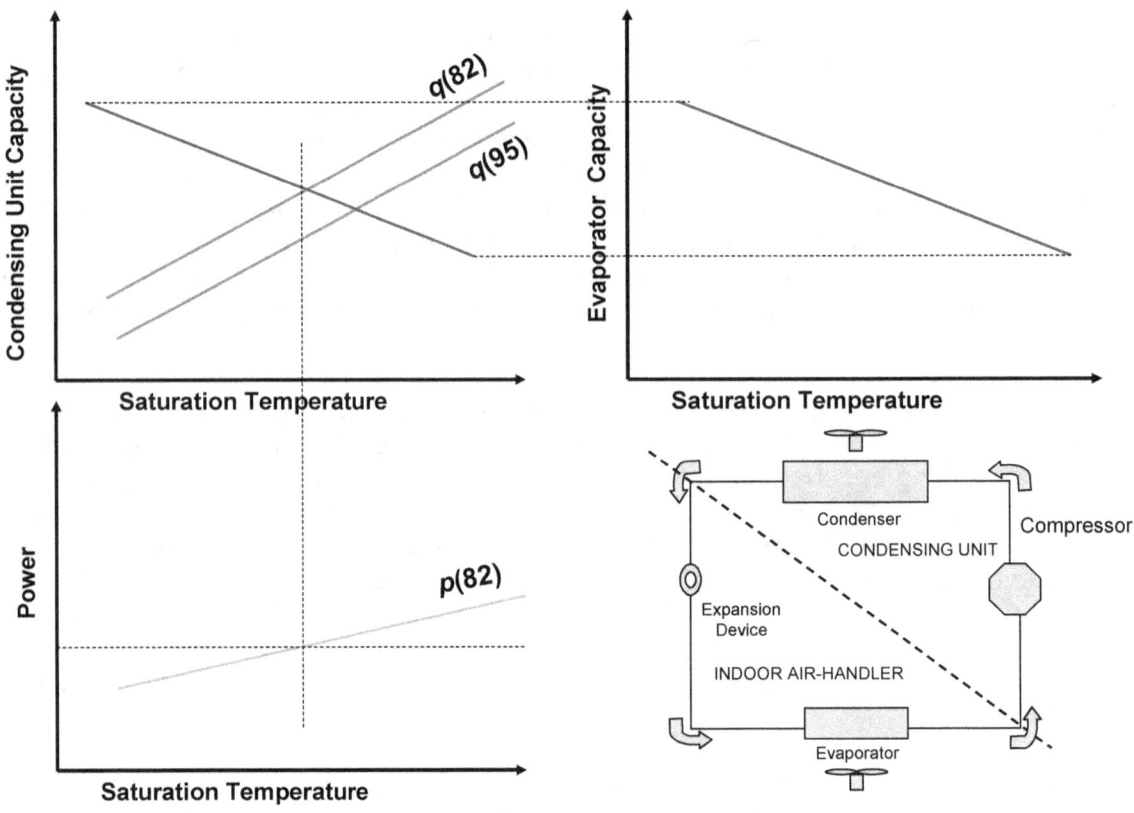

Figure 1.1: Graphical illustration of the single-speed linear fit rating procedure

Two-speed and variable-speed compressor outdoor units may be paired with different indoor units just as single-speed outdoor units are paired with various indoor units. Calculating SEER for modulating equipment is based on a temperature bin method and requires a larger number of test points. Table 1.1 shows the required wet coil, steady-state tests for single-speed, two-speed, and variable-speed equipment as stated in the DOE test requirements (AHRI 2008).

The linear fit method for two-speed mixed systems is graphically depicted in Figure 1.2. Linear fits are required to describe the matched system's cooling capacity and outdoor unit power. For the purposes of explaining the proposed procedure for modulating equipment, this section will focus on two-speed systems, but the discussed concepts still apply to variable-speed equipment. The proposed procedure requires that the rater have the matched system ratings along with the linear fits for power and capacity for the quantities in Table 1.2. Table 1.2 also shows what should be included in all submittals to the AHRI database; these quantities are necessary to bring consistency with single-speed submittals and to allow linear fit comparisons by ICMs.

Table 1.1: Required steady-state, wet coil tests for single-, two-, and variable-speed compressor systems

Indoor/Outdoor Dry Bulb Temperature (°F)	Indoor Air Volume Rate	Compressor Speed	Letter Designations		
			Single-Speed Compressor	Two-Speed Compressor	Variable-Speed Compressor
80/95	Certified Max	Max	A	A_2	A_2
80/95	Min	Min			
80/82	Certified Max	Max	B	B_2	B_2
80/82	Min	Min		B_1	B_1
80/87	Intermediate	Intermediate			E_V
80/67	Min	Min		F_1	F_1

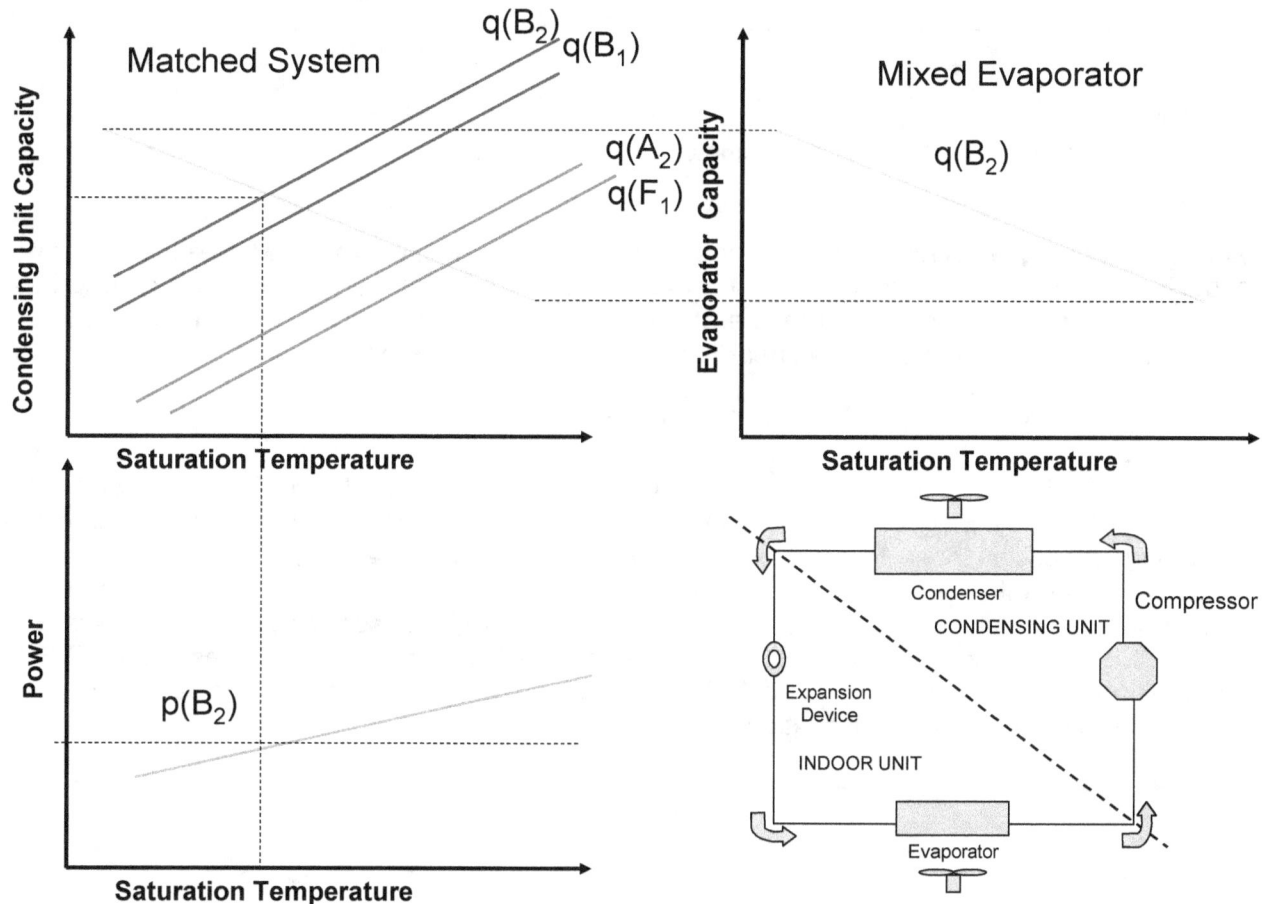

Figure 1.2: Graphical illustration of the two-speed linear fit rating method (determination of $p(B_2)$ is shown as an example)

Table 1.2: Data submitted (*and data that should be added to submittals*) by OEMs to AHRI for two-speed and variable-speed compressor condensing units

	Two-speed				
	Capacity, Power	Capacity, Power			
A_2	Slope	Intercept	T_{suph} *refrigerant*	Indoor airflow	T_{LIQ} refrigerant
B_2	Slope	Intercept	T_{suph} *refrigerant*	*Indoor airflow*	T_{LIQ} refrigerant
B_1	*Slope*	*Intercept*	T_{suph} *refrigerant*	*Indoor airflow*	T_{LIQ} *refrigerant*
F_1	*Slope*	*Intercept*	T_{suph} *refrigerant*	*Indoor airflow*	T_{LIQ} *refrigerant*
	Variable-speed				
	Capacity, Power	Capacity, Power			
A_2	Slope	Intercept	T_{suph} *refrigerant*	Indoor airflow	T_{LIQ} refrigerant
B_2	Slope	Intercept	T_{suph} *refrigerant*	*Indoor airflow*	T_{LIQ} *refrigerant*
B_1	*Slope*	*Intercept*	T_{suph} *refrigerant*	*Indoor airflow*	T_{LIQ} *refrigerant*
E_V	*Slope*	*Intercept*	T_{suph} *refrigerant*	*Indoor airflow*	T_{LIQ} *refrigerant*
F_1	*Slope*	*Intercept*	T_{suph} *refrigerant*	*Indoor airflow*	T_{LIQ} *refrigerant*

1) Italicized and blue-font entries are not currently included in submittals to AHRI, but they need to be included to bring consistency to the AHRI database

When developing a rating for a two-speed mixed air conditioner, the ICM needs evaporator capacity as a function of evaporator exit refrigerant saturation temperature at a fixed superheat and a fixed refrigerant liquid inlet temperature at the expansion device. Or in other words, the ICM needs an evaporator capacity linear fit as described in Equation 1.3.

$$q(A_2, B_2, ...F_1) = \text{Slope} \cdot T_{evap} + \text{Intercept} \qquad 1.3$$

Calculating SEER involves taking the ratio of the sum of the building loads, BL(j), divided by the sum of input energy, E(j), for *j* bins as shown in Equation 1.4, where E(j) terms include the effect of cycling losses for those temperature bins where the system capacity exceeds the building load. Since the information on the cyclic degradation coefficient is not available, Equation 1.4 cannot be used. Instead we may use an approach similar to that used for single-speed systems where $SEER_{mixed}$ is derived from $SEER_{matched}$ by scaling it with the ratio of corresponding EERs and multiplying by F_{exp} (see Equation 1.2). For multi-speed mixed systems the rating equation for SEER will take the form of Equation 1.5.

$$SEER = \frac{\sum BL(j)}{\sum E(j)} \qquad 1.4$$

$$SEER_{mixed} = SEER_{matched} \frac{\sum EER_{j,\,mixed}}{\sum EER_{j,\,matched}} F_{exp} \qquad 1.5$$

The EERs of the mixed and matched systems would be calculated from the linear fits for capacity and power. The calculation of SEER for mixed variable-speed equipment is more cumbersome than for mixed two-speed, but it will still follow the concept given by Equations 1.4 and 1.5.

For mixed systems with a variable-speed compressor, an additional complication and effort will be required over that for two-speed systems because of the intermediate test point. The OEMs provide linear fits for capacity and power at all mandatory test conditions as shown in AHRI 210/240-2008.

The goal of this study was to evaluate the practicality and accuracy of the linear fit method through its application to two-speed mixed systems. In this effort, NIST assumed the role of an evaporator manufacturer and developed cooling capacity lines for two mixed evaporator coil blowers by testing them with a water-cooled condensing unit. To remove any doubt in the CD unit linear fits, NIST also determined linear fits for the CD unit capacity and power by testing the condensing unit with a water-heated evaporator. NIST then developed mixed system ratings based upon the linear fits of the various components and compared the bin-method SEER calculation to a matched system scaled SEER calculation method (illustrated by Equation 1.5).

2: DESCRIPTION OF EVAPORATORS AND MATCHED SYSTEM CONDENSING UNIT

Table 2.1 shows basic information on the tested evaporators. All evaporators were of the finned-tube design. Appendix A presents detailed design data, circuitry configuration, and pictures of the coil blowers and condensing unit. All of the evaporators tested were equipped with a fan and required indoor fan power measurement.

Table 2.1: Evaporator descriptions

Coil Designation	Coil Configuration	AHRI Type	Airflow Direction	Tube Outside Diameter	Expansion Device	Refrigerant
Matched	A	RCU-A-CB	Upflow	9.5 mm (0.375 in)	TXV	R410A
Mixed #1	A	RCU-A-CB	Upflow	9.5 mm (0.375 in)	TXV	R410A
Mixed #2	Inclined Slab	SDHV-RCU-A-CB	Horizontal	9.5 mm (0.375 in)	TXV	R410A

The air-cooled, matched system condensing unit had a two-speed compressor and variable-speed fan. Ratings for this condensing unit with its matched indoor air handler and the first mixed air handler (mixed #1 indoor air handler) are given in Table 2.2. The mixed system ratings for the second air handler (mixed #2 indoor air handler) with this condensing unit were not available, but ratings for this evaporator with a single-speed condensing unit are given.

Table 2.2: Matched and mixed system AHRI directory ratings

System Designation	AHRI Type	Capacity, kW (Btu/h)	EER(A_2) (Btu/Wh)	SEER
Matched	RCU-A-CB	11.13 (38000)	14.60	20.0
Mixed #1	RCU-A-CB	10.60 (36200)	13.00	17.5
Mixed #2	SDHV-RCU-A-CB	10.43 (35600)	9.65	11.0

3: EXPERIMENTAL METHOD

3.1: Experimental setup

Figure 3.1.1 shows the experimental setup, and Appendix B shows detailed pictures of the water cooled condensing unit. The evaporator being tested was installed in the indoor environmental chamber. Air was pulled through the evaporator by a centrifugal fan located at the outlet of the nozzle chamber ductwork. The adjacent outdoor chamber housed the air-cooled condensing unit for system tests or the water-cooled condensing unit and the laboratory water-chiller for evaporator tests.

For evaporator tests with the water-cooled condensing unit, the water chiller control system manipulated the temperature and mass flow rate of the water delivered to the condensing unit. The chiller rejected heat to the in-house chilled water loop. Heat rejection was to water and did not require maintaining the outdoor chamber conditions.

The installation of the evaporator and test instrumentation conformed to ASHRAE Standard 37 and AHRI Standard 210/240. We used the air enthalpy method for the primary measurement of the evaporator capacity with the refrigerant enthalpy method providing the secondary measurement. Air dew-point temperature was measured at the inlet of the evaporator ductwork and in the ductwork after the evaporator and several mixers. Twenty-five node thermocouple grids, located on each side of the evaporator, were used to verify that the air was well mixed at each point. A 25-junction thermopile measured the air temperature change across the evaporator. Barometric pressure, evaporator air pressure drop, air temperature and pressure drop in the nozzle, and nozzle temperature were used along with the dew-point measurements to establish the thermodynamic state of the air. The refrigerant enthalpy method was the secondary measurement of evaporator capacity and required measurement of the evaporator inlet and exit refrigerant temperatures and pressures in addition to mass flow rate. The agreement between the air-side and refrigerant-side methods was always within 6 %.

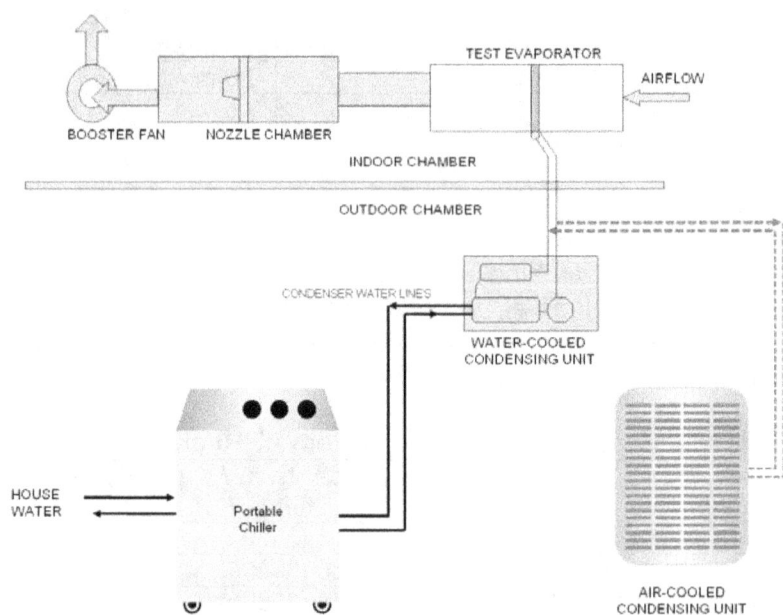

Figure 3.1.1: Experimental setup for evaporator and system testing

3.2: Data acquisition and measurement uncertainty

The measurements consisted of temperature, pressure, pressure difference, temperature difference, dew-point temperature, fan amps, fan volts, and fan power. Table 3.2.1 lists the measured quantities and their uncertainties for a 95 % confidence limit (two sigma on the mean value) (Taylor and Kuyatt 1994).

Table 3.2.1: Measurement uncertainty

Quantity	Range	Uncertainty
Pressure	0 kPa to 3447 kPa (0 psia to 500 psia)	±3.4 kPa (±0.5 psi)
Temperature	-26.1 °C to 93.3 °C (-15 °F to 200 °F)	±0.3 °C (±0.5 °F)
Temperature Difference	0 °C to 27.8 °C (0 °F to 50 °F)	±0.3 °C (±0.5 °F)
Barometric Pressure	0 mm Hg to 1270 mm Hg (0 in Hg to 50 in Hg)	±0.34 mm Hg (±0.0135 in Hg)
Dew-point Temperature	0 °C to 50 °C (32 °F to 122 °F)	±0.2 °C (±0.4 °F)
Pressure Difference	0 Pa to 374 Pa (0 in H_2O to 1.5 in H_2O)	±3.8 Pa (±0.02 in H_2O)
Mass Flow	0 kg/h to 544.3 kg/h (0 lb/h to 1200.0 lb/h)	±1 %
Evaporator Capacity	5.56 kW to 14.4 kW (19 kBtu/h to 49 kBtu/h)	±3 % to ±7 %

4: TWO-SPEED MATCHED AND MIXED SYSTEM TESTS

Tables 4.1 through 4.3 list the test conditions and test results for the matched and mixed systems.

Table 4.1: Matched system tests

Test Designation	Indoor Airflow, scfm	Average Evaporator Exit Refrigerant Saturation Temperature, °F	Average Refrigerant Liquid Line Temperature, °F	Total Capacity, Btu/h	Sensible Heat Ratio	EER Btu/kWh
A$_2$	1240	52.7	95.8	36081	0.74	12.96
A$_1$	942	54.1	95.1	25307	0.77	14.85
B$_2$	1240	51.4	82.9	38897	0.71	15.73
B$_1$	940	52.9	82.9	27484	0.74	19.48

Table 4.2: Mixed system #1 tests

Test Designation	Indoor Airflow, scfm	Average Evaporator Exit Refrigerant Saturation Temperature, °F	Average Refrigerant Liquid Line Temperature, °F	Total Capacity, Btu/h	Sensible Heat Ratio	EER Btu/kWh
A$_2$	1215	47.1	96.5	31967	0.78	11.27
A$_1$	965	49.1	96.2	22300	0.82	12.11
B$_2$	1222	46.4	83.9	34905	0.75	13.78
B$_1$	971	48.5	83.7	24307	0.80	15.46

Table 4.3: Mixed system #2 tests

Test Designation	Indoor Airflow, scfm	Average Evaporator Exit Refrigerant Saturation Temperature, °F	Average Refrigerant Liquid Line Temperature, °F	Total Capacity, Btu/h	Sensible Heat Ratio	EER Btu/kWh
A$_2$	750	46.3	96.6	29981	0.63	9.87
A$_1$	753	51.4	96.7	22263	0.72	10.23
B$_2$	760	45.2	84.1	32652	0.62	11.99
B$_1$	753	50.1	84.2	23864	0.70	12.53

5: MATCHED COIL TESTS

The matched system's air handler was attached to a water-cooled condensing unit and tested over a range of evaporator exit saturation temperatures, evaporator exit superheats and refrigerant liquid inlet temperatures as shown in Table 5.1. These tests allowed linear fits to be developed for cooling capacity as a function of evaporator exit refrigerant saturation temperature at a constant superheat for the various liquid temperatures corresponding to the standard test conditions. Since the matched system was a two-speed system, the E_v test was not required, but data was taken for an inlet refrigerant liquid temperature near 87 °F to explore the effects of liquid refrigerant temperature on cooling capacity and to illustrate the usefullness of the linear fit method for variable-speed equipment.

Table 5.1: Matched coil performance summary

Test	Evaporator Exit Saturation Temperature w/ Superheat, Low – High, °F[1]	Coil Only Cooling Capacity, High – Low, Btu/h [2]	Refrigerant Liquid Temperature, Low – High, °F	Range of Coil Sensible Heat Ratio
A_2	(47.8, 19.8)-(55.6, 5.8)	25921 – 21641	94.7 – 95.4	0.93 – 0.78
A_1	(46.0, 20.0) – (53.5, 9.6)	30188 – 21386	94.5 – 95.5	0.85 – 0.71
B_2	(47.9, 20.6) – (54.0, 10.2)	34455 – 24871	81.9 – 82.5	0.89 – 0.74
B_1	(47.0, 9.8) – (52.5, 10.1)	33712 – 23526	81.9 – 82.4	0.80 – 0.69
E_v	(47.2, 10.7) – (52.7, 10.2)	32079 – 22239	87.2 – 87.5	0.83 – 0.70
F_1	(45.9, 10.5) – (52.2, 10.4)	36673 – 25697	67.1 – 67.4	0.78 – 0.67

[1] – Evaporator exit refrigerant saturation temperature and superheat (T_{evap}, T_{suph})
[2] – Capacity at the temperature conditions listed in column 2

5.1: Matched coil test results at A_2, A_1, B_2, B_1, E_v, and F_1 conditions

Table 5.1.1 shows the linear fits for the matched coil capacity as a function of evaporator exit saturation temperature at various superheats. It would have been better to have more than three points to do a linear fit, but time was a limiting factor and thus a visual inspection of the trends seen in the linear fits with varying superheats is more indicative of the relationships than a purely numerical analysis would indicate. The uncertainty in the capacity measurement at a given evaporator temperature is at least ±3 %; this is neglecting the uncertainty of measuring the evaporator temperature and superheat which also adds uncertainty to the capacity determination and the linear fit. Visually inserting these error bars onto the data points and extending all possible lines through the resulting range of points is one technique for seeing similarities in the linear fits that may be confounded by comparing a strict linear fit to the available data. For example, the B_1 and A_1 tests in Figure 5.1.1 at a superheat of 10 °F appear

11

to have very similar slopes, but they are not numerically equal in Table 5.1.1, yet visually they appear equal and certainly the uncertainty in their slopes would overlap due to only three points being used in the linear fit (see Payne and Domanski 2006 for a more detailed uncertainty analysis of linear fits).

It is interesting to note the similarities in slopes between all tests at high airflow rates and low airflow rates; B_2 and A_2 have very similar slopes just as B_1, A_1, and E_v show similar slopes. Even F_1 has a similar slope to the other low airflow tests, but it appears that the negative approach temperature (67 $^\circ$F - 80 $^\circ$F = -13 $^\circ$F) differentiates the F_1 test from the other low airflow rate tests with positive approach temperatures for this coil.

Table 5.1.1: Linear fits of matched coil-only capacity as a function of evaporator exit refrigerant saturation temperature (does not include fan heat)

Test	Number of Points in linear fit	Slope, Btu/(h$^\circ$F)	Intercept, Btu/h	Pearson's Correlation Coefficient, R^2	Airflow, scfm	Average Blower Power (W)[1]
A_2 (Tsuph=10 $^\circ$F)	4	-2703.7	171680	0.995	1225	274
A_2 (Tsuph=5 $^\circ$F)	3	-3428.8	212735	0.976	1231	280
A_2 (Tsuph=20 $^\circ$F)	3	-2634.2	160962	0.999	1232	284
A_1 (Tsuph=10 $^\circ$F)	3	-1814.5	118362	0.999	948	121
A_1 (Tsuph=20 $^\circ$F)	3	-1835.2	114263	0.988	945	121
B_2 (Tsuph=10 $^\circ$F)	3	-2774.5	174701	0.999	1226	278
B_2 (Tsuph=20 $^\circ$F)	3	-2119.2	135725	0.999	1229	279
B_1 (Tsuph=10 $^\circ$F)	6	-1839.4	120219	0.996	946	121
E_v (Tsuph=10 $^\circ$F)	3	-1769.0	115665	0.998	947	119
F_1 (Tsuph=10 $^\circ$F)	4	-1725.9	116067	0.997	950	121

[1]- Total static pressure drop seen across air handler was 60 Pa (0.24 inches of water gage)

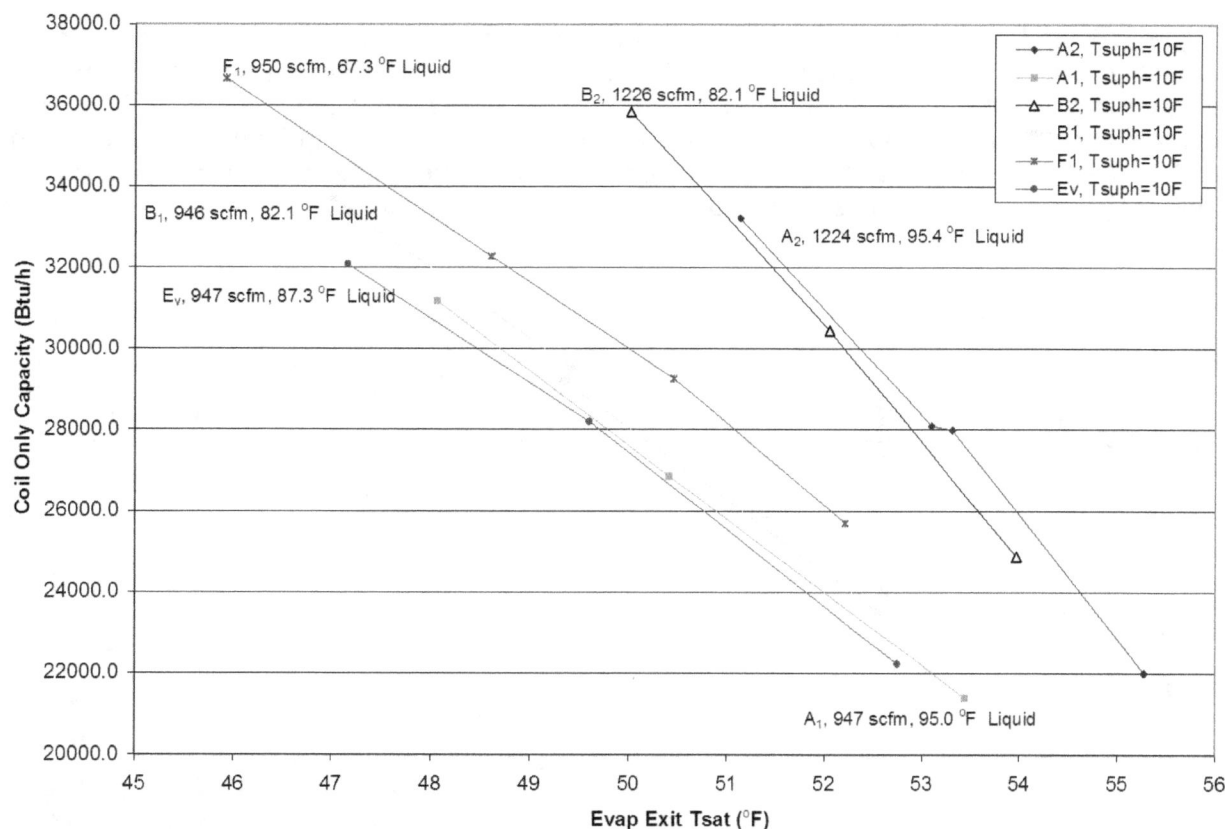

Figure 5.1.1: Matched coil total cooling capacity (fan heat not included, E_v is low compressor speed)

13

Figure 5.1.2 adds a higher superheat to the data presented in Figure 5.1.1. Visually, the effect of the higher superheat is a lowering of the linear intercept with almost constant slopes for a given test condition; tests at comparable liquid temperatures but different superheats appear to differ in capacity by a constant offset. This trend is illustrated again in Figure 5.1.3 for the high airflow case with near 95 $^{\circ}$F liquid refrigerant temperatures; three superheats are shown indicating almost equal slopes with a constant offset in cooling capacity.

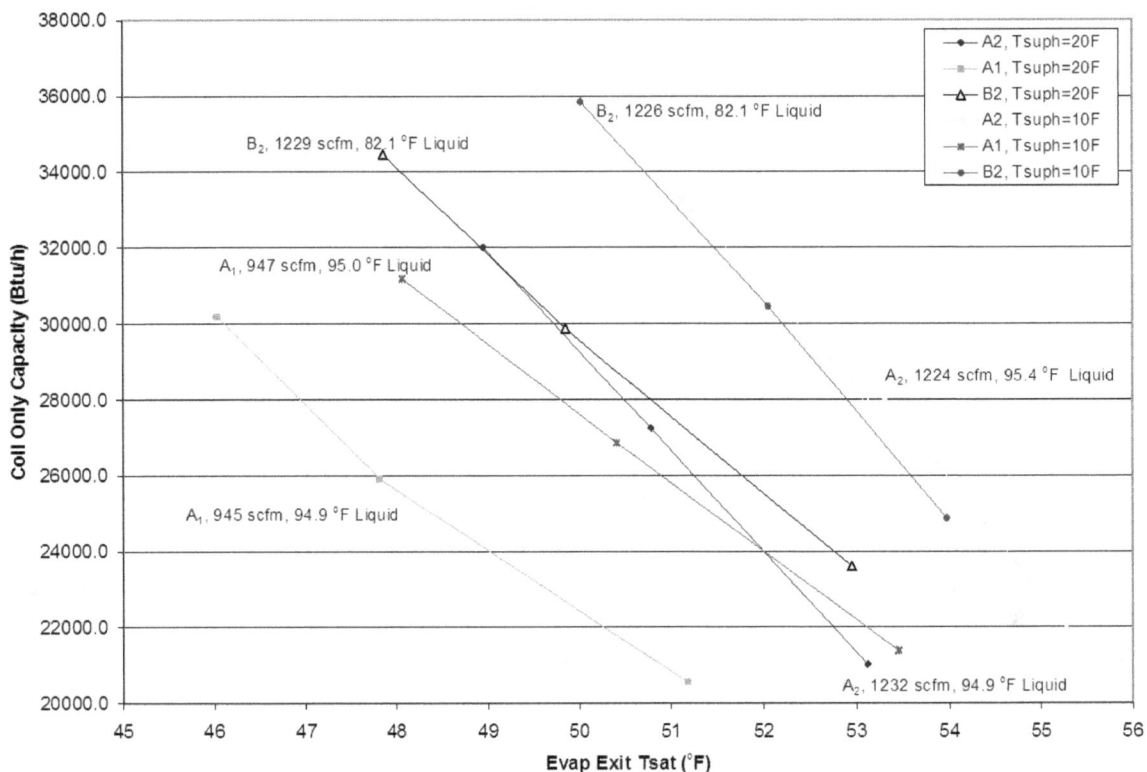

Figure 5.1.2: Matched coil capacity at two different superheats

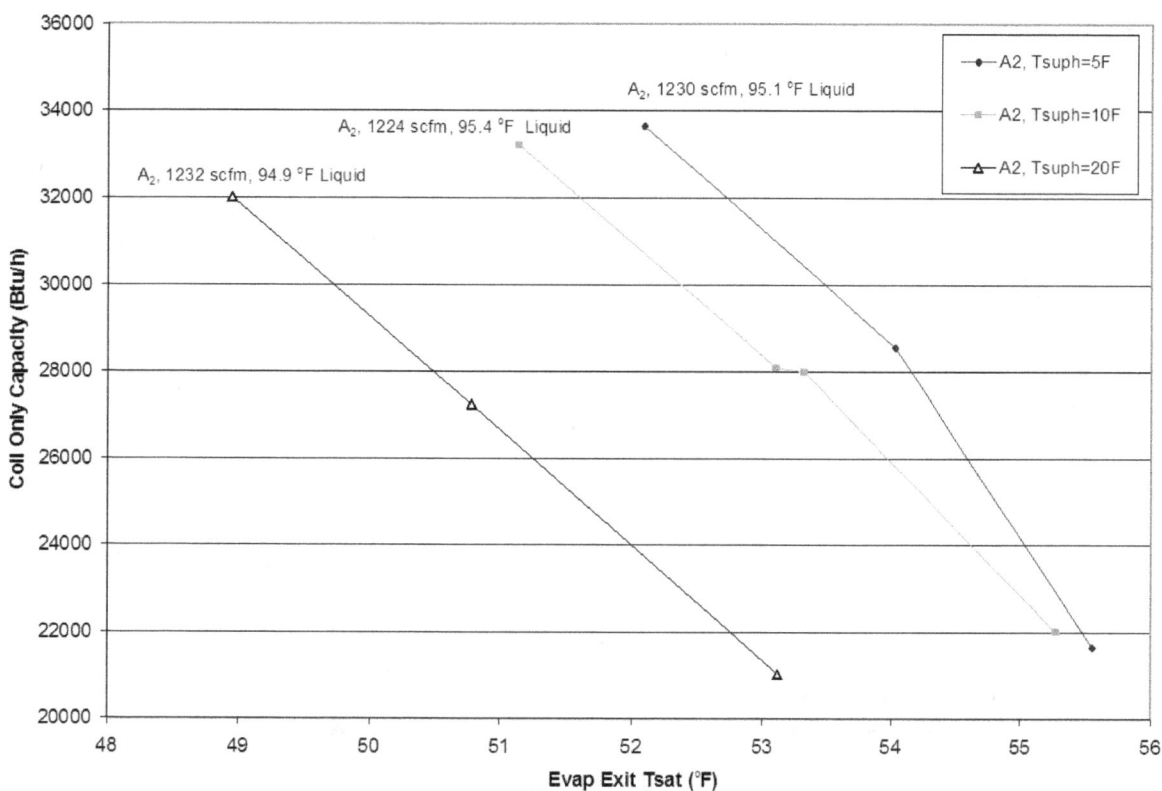

Figure 5.1.3: Coil capacity at A_2 conditions for three refrigerant superheats

5.2: Matched coil airflow specific cooling capacity

Figure 5.2.1 shows the coil-only cooling capacity divided by the standard airflow rate as a function of the evaporator exit refrigerant saturation temperature. All refrigerant liquid temperatures and airflow rates are presented in this figure for a constant evaporator exit refrigerant superheat. Sensible and latent capacity lines are broken out of the total capacity to show dehumidification performance as a function of saturation temperature.

Figure 5.2.2 shows the airflow specific capacity for the matched coil at different superheats. As seen in the previous linear fits, the F_1 test points do not strictly group with the other total capacity line, but the three points are within +5.5 % of the line. All liquid temperatures are represented and seem to produce a weak effect on capacity.

15

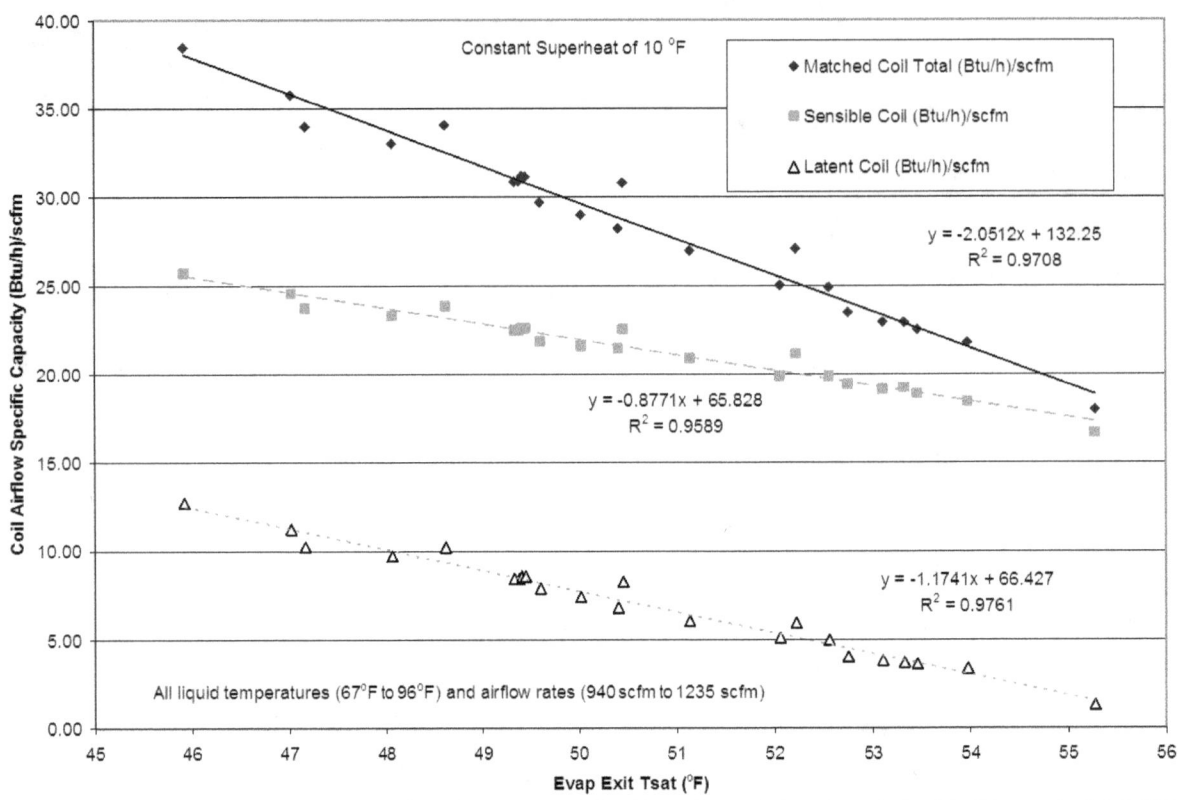

Figure 5.2.1: Linear fits to total, sensible, and latent cooling capacity per unit airflow rate for all tests combined at a superheat of 10.0 °F (coil-only capacity with no accounting for fan heat)

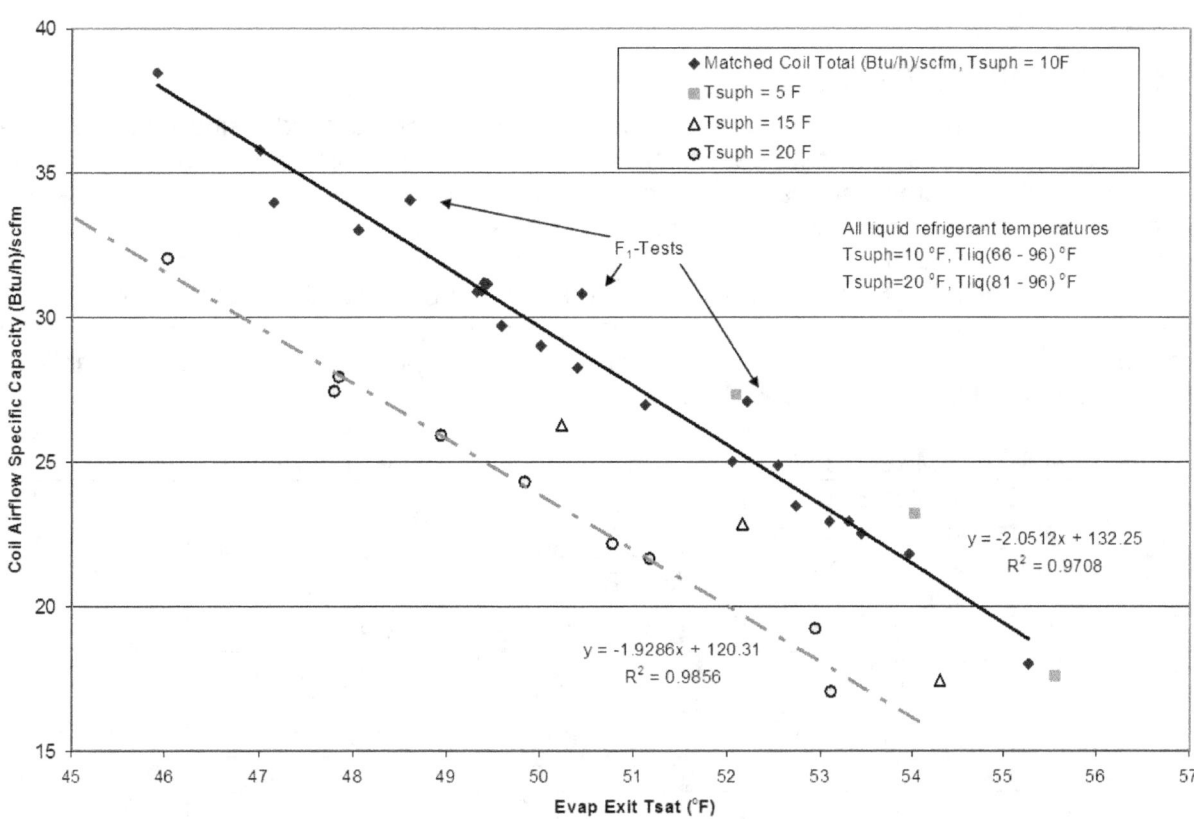

Figure 5.2.2: Matched coil total cooling capacity per unit airflow rate at different superheats
(coil-only capacity with no accounting for fan heat)

6: MIXED COIL #1 TESTS

The mixed #1 system's air handler was attached to a water-cooled condensing unit and tested over a range of evaporator exit saturation temperatures, evaporator exit superheats and refrigerant liquid inlet temperatures as shown in Table 6.1. These tests allowed linear fits to be developed for cooling capacity as a function of evaporator exit refrigerant saturation temperature at a constant superheat at the various liquid temperatures corresponding to the standard test conditions. Since the matched system was a two-speed system, the E_v test was not required, but data was taken for inlet refrigerant liquid temperature near 87 °F to explore the effects of liquid refrigerant temperature on cooling capacity and to illustrate the applicability of the linear fit method for variable-speed equipment.

Table 6.1: Mixed coil #1 performance at various evaporator temperatures

Test	Evaporator Exit Saturation Temperature and Superheat, Low – High, °F [1]	Coil Only Cooling Capacity, Low – High, Btu/h [2]	Refrigerant Liquid Temperature, Low – High, °F	Range of Coil Sensible Heat Ratio
A_2	(45.0, 10.0) – (53.2, 9.8)	34755 – 17130	94.8 – 105.4	0.99 – 0.76
A_1	(45.8, 10.2) – (54.0, 10.0)	27682 – 13501	94.8 – 105.4	0.99 – 0.76
B_2	(46.1, 9.8) – (52.8, 10.4)	33144 – 16914	82.1 – 88.7	0.99 – 0.77
B_1	(43.8, 10.2) – (51.4, 10.4)	31791 – 16432	82.0 – 88.6	0.94 – 0.71
E_v	(43.8, 10.2) – (51.4, 10.4)	31791 – 16432	84.5 – 88.6	0.94 – 0.71
F_1	(42.6, 10.1) – (48.0, 10.2)	34114 – 22999	66.6 – 73.3	0.83 – 0.69

[1] – Evaporator exit refrigerant satuation temperature and superheat (T_{evap}, T_{suph})
[2] – Capacity at the temperature conditions listed in column 2

6.1: Mixed coil #1 linear fits at A_2, A_1, B_2, B_1, E_v, and F_1 conditions

Table 6.1.1 shows the linear fits for the mixed #1 coil at the various standard test conditions and a constant superheat. Liquid refrigerant temperature entering the expansion valve was varied around the outdoor air temperature corresponding to the given test condition; if the coil had been connected to an air-cooled condensing unit, the refrigerant liquid temperature would be close to or higher than the outdoor air temperature.

As seen with the matched coil, capacity slopes at high airflow rates and low airflow rates are similar (Figure 6.1.1). Figures 6.1.2 through 6.1.6 show the weak effects of different liquid refrigerant inlet temperatures on coil capacity.

Table 6.1.1: Linear fits of mixed coil #1, coil-only capacity as a function of evaporator exit refrigerant saturation temperature (does not include fan heat)

Test	Number of Points in linear fit[1]	Slope, Btu/(h°F)	Intercept, Btu/h	Pearson's Correlation Coefficient, R^2	Airflow, scfm	Average Blower Power, W[2]
A_2 (Tsuph=10 °F)	12	-2072.6	127404	0.994	1210	378
A_1 (Tsuph=10 °F)	11	-1771.6	108723	0.995	964	225
B_2 (Tsuph=10 °F)	9	-2353.5	141020	0.989	1213	374
B_1 (Tsuph=10 °F)	9	-1971.9	118075	0.998	967	228
E_v (Tsuph=10 °F)	6	-2001.3	119593	0.999	967	232
F_1 (Tsuph=10 °F)	10	-2183.1	127818	0.998	967	226

[1]- Includes all refrigerant liquid temperatures
[2]- External static pressure drop seen across air handler was (0.22 to 0.24) inches water gage

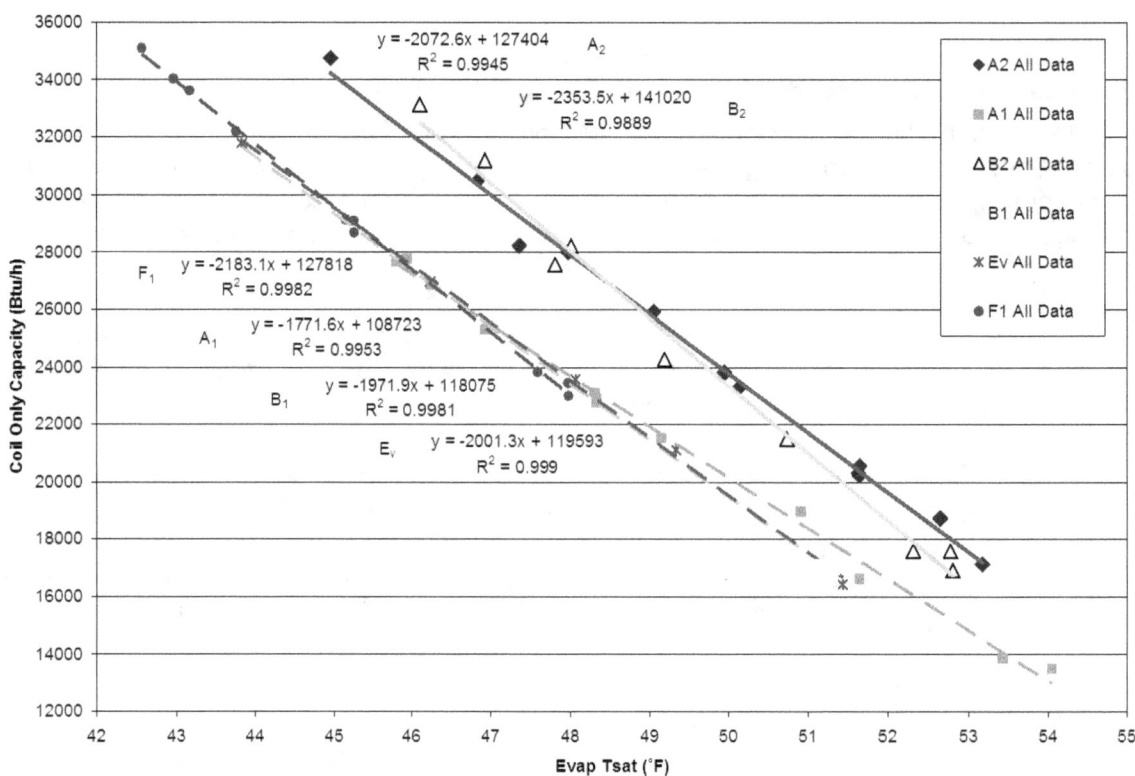

Figure 6.1.1: Mixed coil #1 coil-only capacity for all test conditions for all liquid refrigerant temperatures and constant superheat of 10 °F

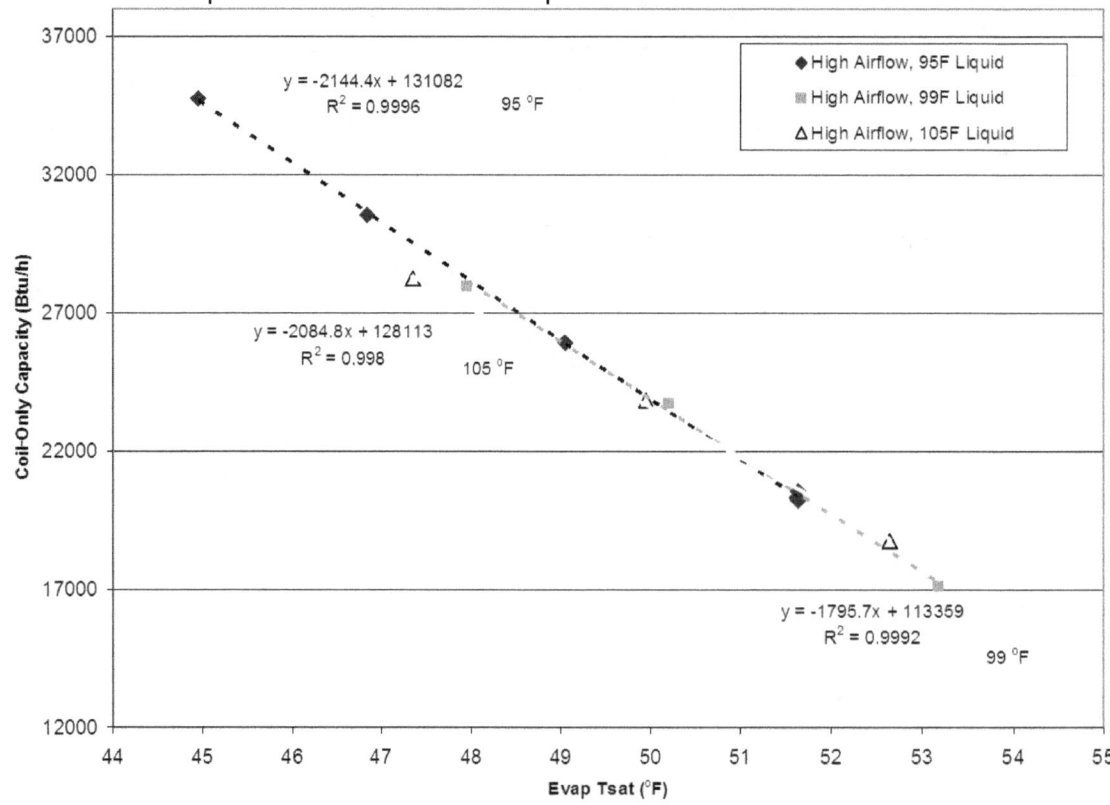

Figure 6.1.2: Mixed coil #1 A_2 coil-only capacity at three different refrigerant liquid temperatures and superheat of 10 °F

20

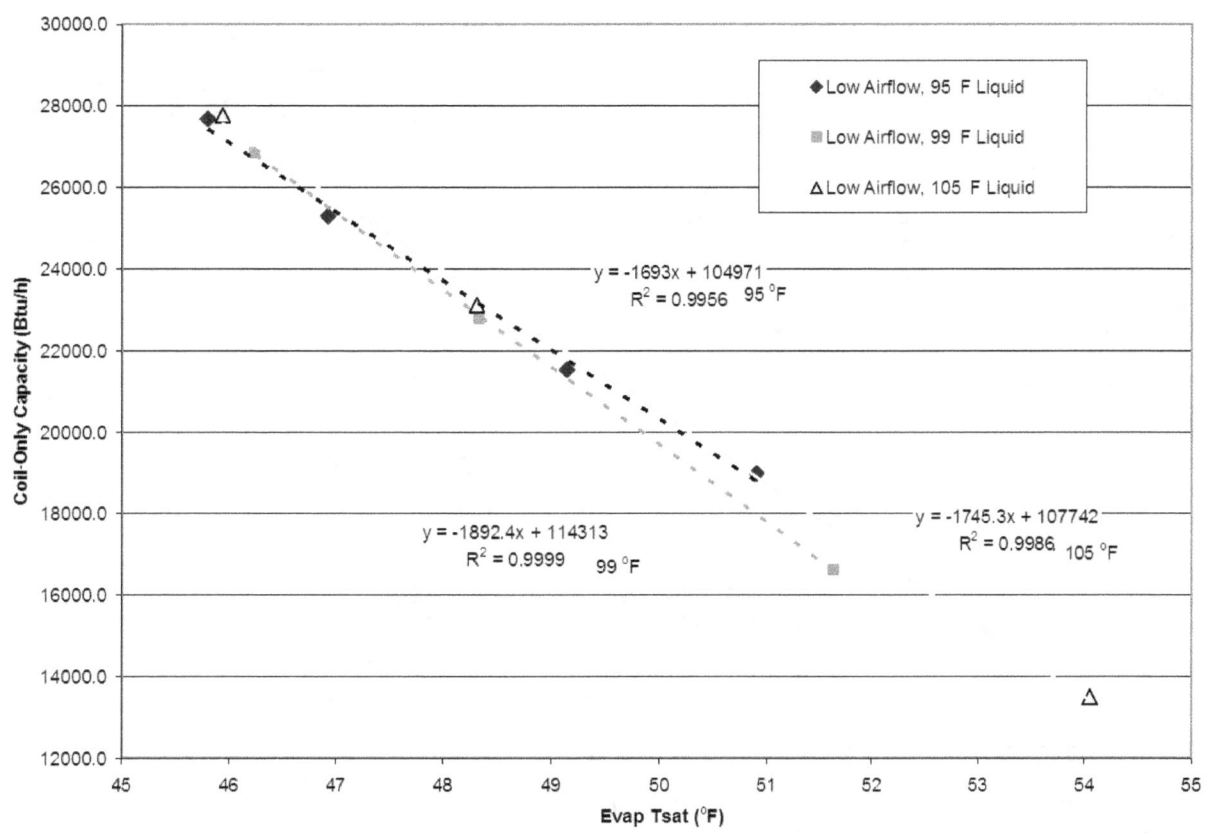

Figure 6.1.3: Mixed coil #1 A_1 coil-only capacity at different refrigerant liquid temperatures

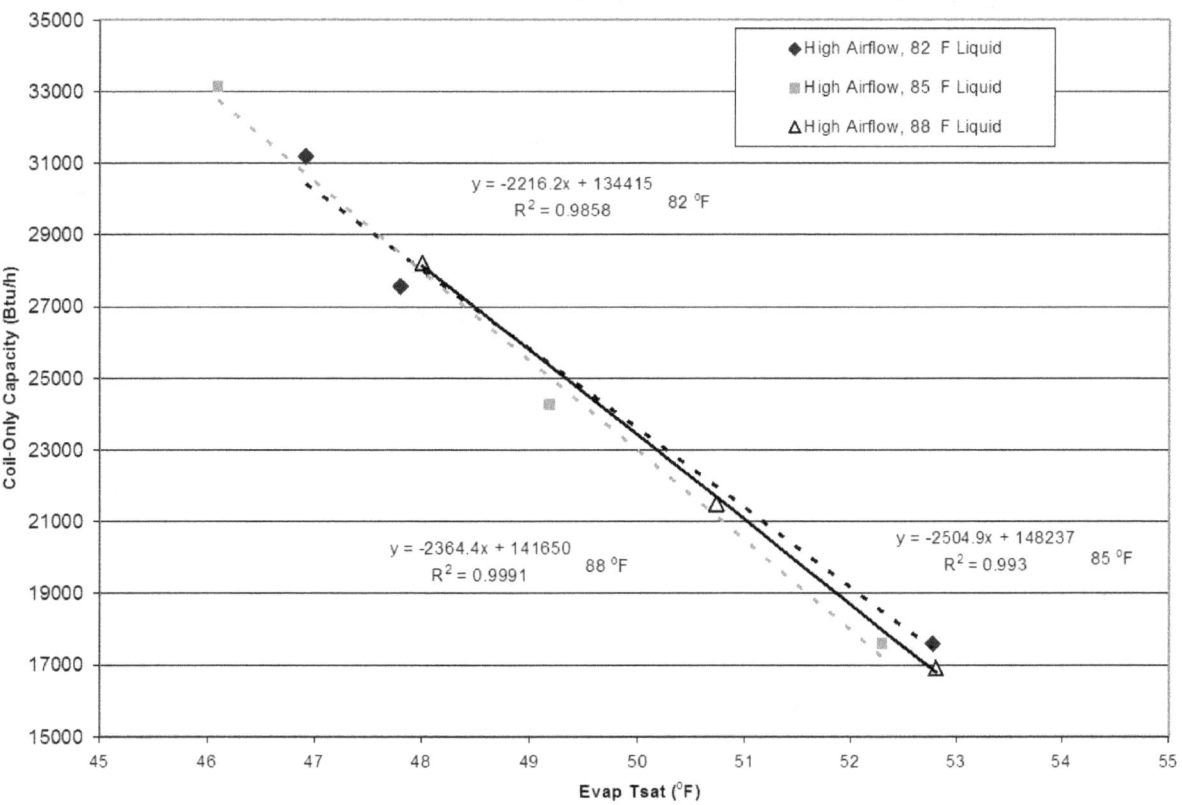

Figure 6.1.4: Mixed coil #1 B_2 coil-only capacity at different refrigerant liquid temperatures

21

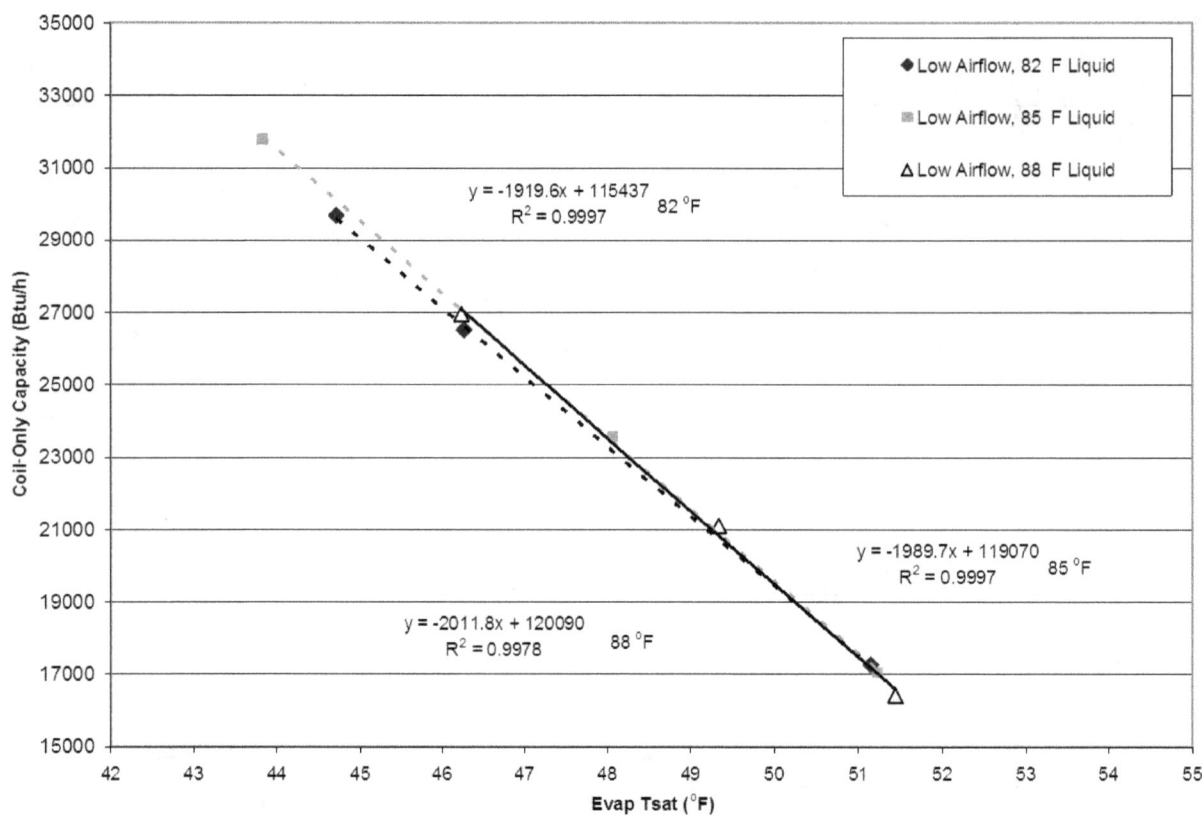

Figure 6.1.5: Mixed coil #1 B_1 and E_v coil-only capacity for different refrigerant liquid temperatures

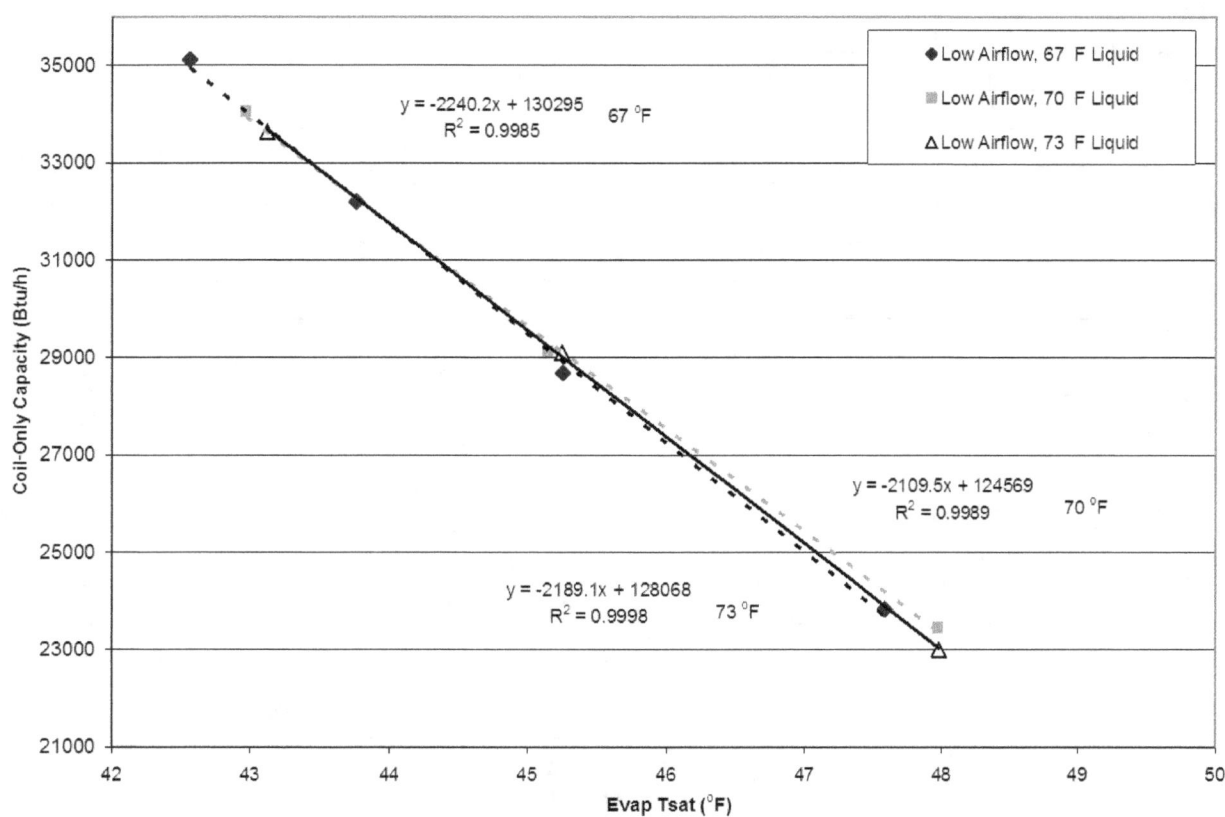

Figure 6.1.6: Mixed coil #1 F_1 coil-only capacity for different refrigerant liquid temperatures

6.2: Mixed coil #1 airflow specific cooling capacity

As seen with the matched coil, mixed #1 coil airflow specific capacity was very linear with evaporator exit refrigerant saturation temperature. Figure 6.2.1 shows total, sensible and latent airflow specific capacity for all liquid temperatures and airflow rates at a constant superheat. The F_1 test does not stand out for mixed #1 coil as it did for the matched coil. This difference may be due to a coil circuiting or coil geometry effect (face velocity, etc.).

23

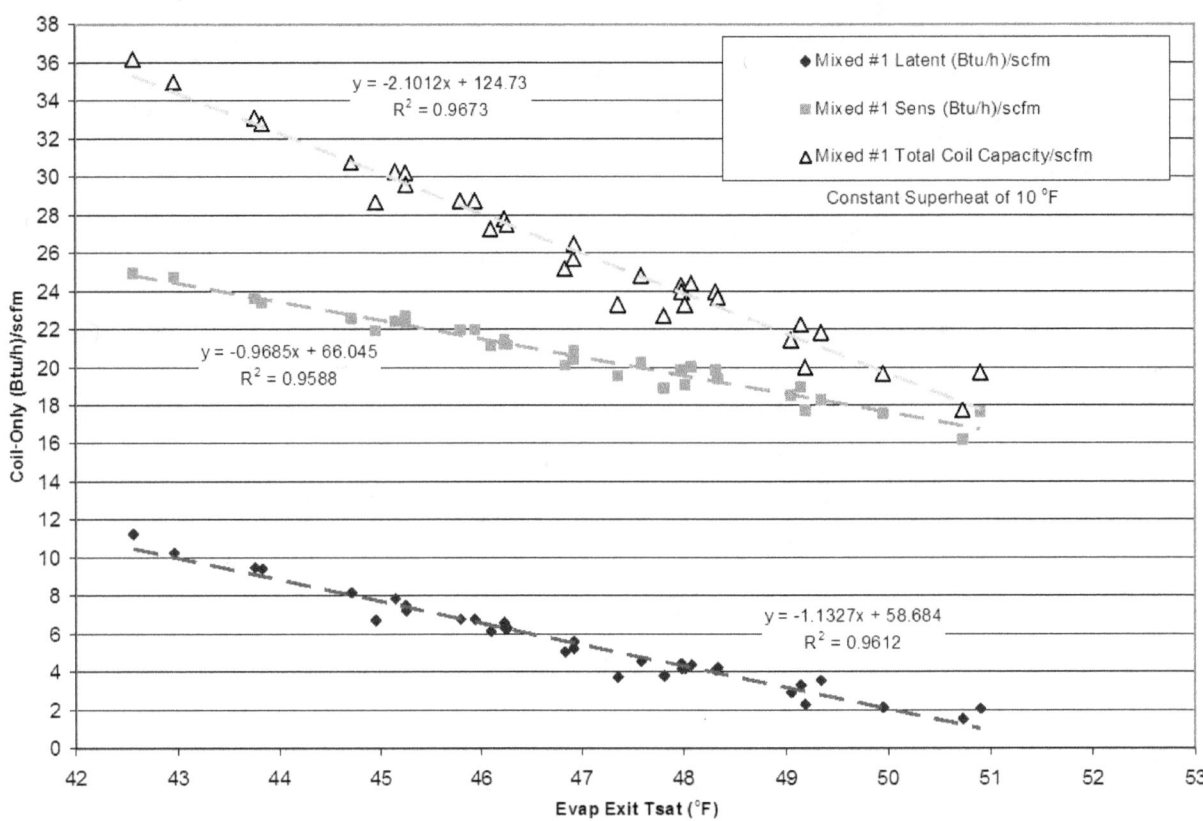

Figure 6.2.1: Mixed coil #1 coil-only capacity per unit airflow rate for all liquid temperatures and a superheat of 10 °F (SHR=1.0 @ 51.8 °F)

7: MIXED COIL #2 TESTS

Table 7.1 shows the range of evaporator temperatures tested and the resulting cooling capacities and sensible heat ratios for all of the tests performed. The mixed #2 coil was part of a small duct, high velocity air handler. In addition to operating at external static pressures greater than 1.2 inH2O, the sensible heat ratios for this air handler were lower than the matched and mixed #1 coils at comparable evaporator saturation temperatures.

Table 7.1: Mixed coil #2 performance at various evaporator temperatures

Test	Evaporator Exit Saturation Temperature w/ Superheat, Low – High, °F [1]	Coil Only Cooling Capacity, Low – High, Btu/h [2]	Refrigerant Liquid Temperature, Low – High, °F	Range of Coil Sensible Heat Ratio
A_2	(47.1, 10.2) – (53.7, 10.0)	31121 – 20214	94.6 – 105.4	0.80 – 0.66
A_1	(44.9, 9.8) –(52.1, 10.4)	28753 – 19487	94.6 – 105.6	0.73 – 0.63
B_2	(46.0, 10.3) – (52.5, 11.6)	33117 – 22755	81.7 – 88.3	0.77 – 0.66
B_1	(44.2, 10.0) – (50.6, 10.5)	30216 – 21453	88.0 – 88.1	0.72 – 0.63
E_v	(44.2, 10.0) – (50.6, 10.5)	30216 – 21453	88.0 – 88.1	0.72 – 0.63
F_1	(41.4, 10.3) – (49.0, 10.5)	33244 – 23758	66.7 – 73.2	0.69 – 0.61

[1] – Evaporator exit refrigerant saturation temperature and superheat (T_{evap}, T_{suph})
[2] – Capacity at the temperature conditions listed in column 2

7.1: Mixed coil #2 linear fits at A_2, A_1, B_2, B_1, E_v, and F_1 conditions

Table 7.1.1 shows the linear fits for the mixed #2 coil at airflow rates and liquid refrigerant temperatures corresponding to the standard test conditions with a constant evaporator exit refrigerant superheat. As shown in Figure 7.1.1 and 7.1.2, liquid refrigerant temperature had a weak effect on coil capacity even for the low temperature liquid refrigerant tests.

Table 7.1.1: Linear fits of mixed coil #2 coil-only capacity as a function of evaporator exit refrigerant saturation temperature (does not include fan heat)

Test	Number of Points in linear fit[1]	Slope, Btu/(h°F)	Intercept, Btu/h	Pearson's Correlation Coefficient, R^2	Airflow, scfm	Average Blower Power, W [2]
A_2 (Tsuph=10 °F)	11	-1631.3	107936	0.99	761	561
A_1 (Tsuph=10 °F)	9	-1394.7	91813	0.99	607	483
B_2 (Tsuph=10 °F)	11	-1717.5	112592	0.99	763	566
B_1 (Tsuph=10 °F)	2	-1366.4	90651	1.0	617	491
E_v (Tsuph=10 °F)	Same as B_1					
F_1 (Tsuph=10 °F)	7	-1336.0	88883	0.99	607	484

[1]- Includes all refrigerant liquid temperatures near the test condition's outdoor air temperature
[2]- Total external static pressure drop seen across air handler was 1.8 inches of water gage or greater

26

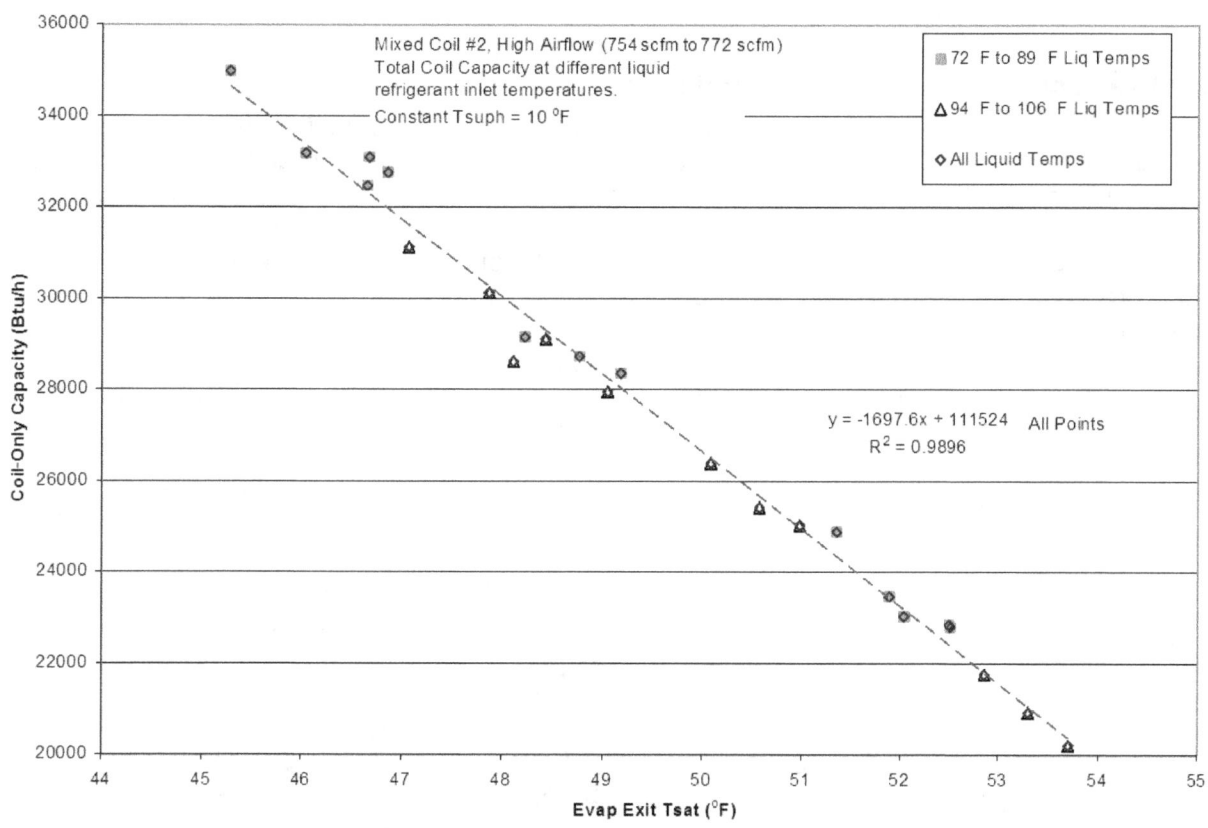

Figure 7.1.1: Mixed coil #2 coil-only capacity, high airflow, A_2 and B_2 conditions

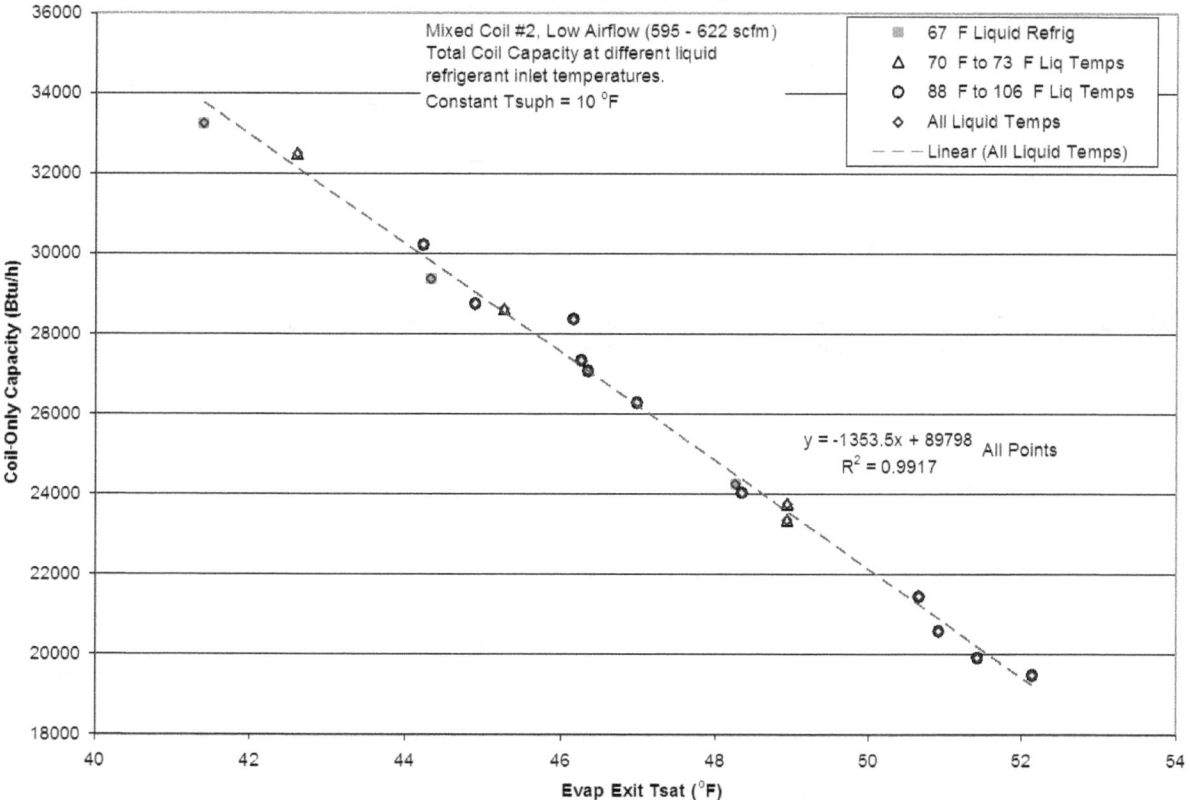

Figure 7.1.2: Mixed coil #2 coil-only capacity, low airflow, A_1, B_1, E_v, and F_1 conditions

7.2: Mixed coil #2 airflow specific cooling capacity

Figure 7.2.1 shows the airflow rate specific capacity for the mixed #2 coil as a function of evaporator exit refrigerant saturation temperature. All approach temperatures are represented well by this linear fit; there is no offset for the F_1 tests as was seen in the matched coil tests.

Figure 7.2.2 shows the effect of different superheats on the airflow specific capacity. The results are very linear at the various superheats; there is only an offset between the various superheats.

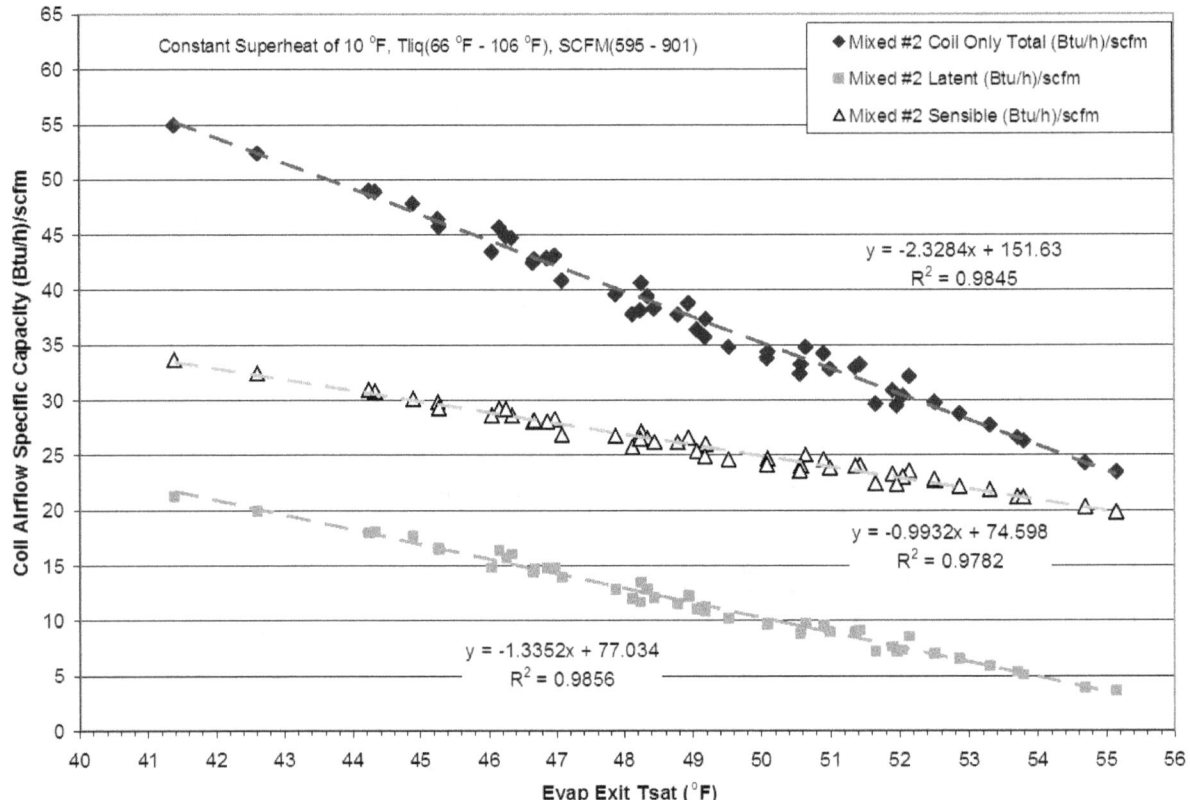

Figure 7.2.1: Mixed #2 coil-only capacity per unit airflow rate for all liquid temperatures and a superheat of 10.0 °F (SHR=1.0 @ 57.7 °F)

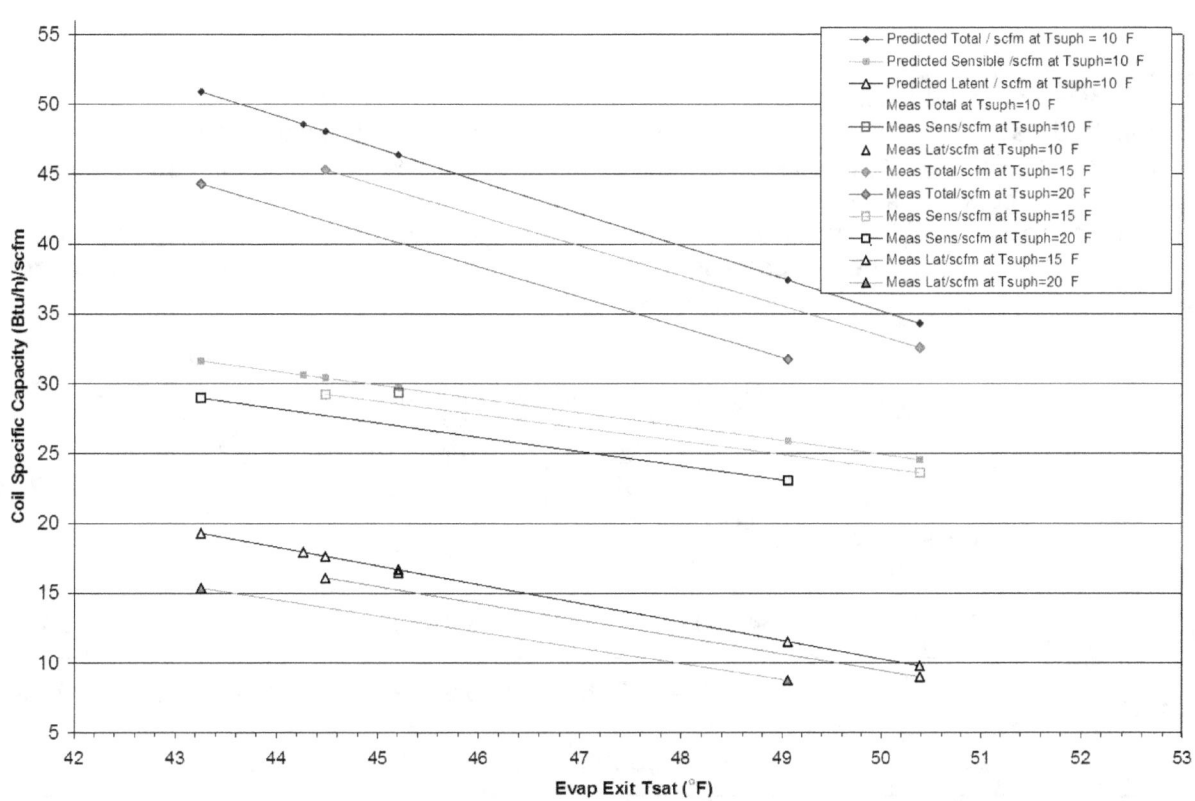

Figure 7.2.2: Mixed #2 coil-only capacity per unit airflow rate at different superheats

29

8: MATCHED CONDENSING UNIT TESTS

The matched system condensing unit was connected to a water-heated evaporator arrangement as shown in Appendix A. The CD unit was located in the outdoor psychrometric chamber and air conditions were established at the various standard test conditions. Table 8.1 shows the range of tests performed with the matched system condensing unit connected to the water-heated evaporator. The Ev test was performed at low compressor speed and low outdoor airflow as established by the CD unit's controls.

Table 8.1: Matched condensing unit capacity

Test	OD Vapor at Service Valve Temperature w/ Superheat, Low – High, °F [1]	Refrig. Side Cooling Capacity, Low – High, Btu/h [2]	Refrigerant Liquid Temperature, Low – High, °F	Refrig. Subcooling at OD Service Valve, Low – High, °F	OD Total Power at Conditions in Col. 2, W
A₂ (Tsuph = 10 °F)	(46.3, 9.7) – (57.6, 10.1)	34330 – 41911	97.7 – 99.4	7.6 – 8.6	2428 – 2558
A₂ (Tsuph = 15 °F)	(43.9, 14.9) – (53.5, 14.8)	33012 – 39386	96.9 – 97.6	8.3 – 9.5	2404 – 2514
A₂ (Tsuph = 20F)	(43.9, 20.0) – (53.3, 20.1)	33108 – 39398	96.6 – 97.0	9.1 – 10.4	2417 – 2518
A₁ (Tsuph = 10 °F)	(47.7, 10.0) – (53.6, 10.2)	23600 – 26760	97.6 – 99.4	4.8 – 7.2	1594 – 1579
A₁ (Tsuph = 15 °F)	(45.5, 14.9) – (57.0, 15.1)	22687 – 28788	96.6 – 98.4	5.3 – 8.8	1601 – 1568
A₁ (Tsuph = 20 °F)	(41.5, 20.1) – (58.5, 20.0)	21105 – 29538	96.5 – 97.9	6.7 – 8.4	1620 – 1558
B₂ (Tsuph = 10 °F)	(46.4, 10.1) – (57.7, 10.1)	37430 – 46090	83.8 – 84.1	10.1 – 12.1	2145 – 2270
B₂ (Tsuph = 15 °F)	(45.1, 15.2) – (56.2, 15.3)	36555 – 44848	83.5 – 83.7	10.2 – 12.2	2132 – 2256
B₂ (Tsuph = 20 °F)	(41.6, 20.2) – (52.9, 20.3)	34343 – 42350	83.1 – 83.3	9.9 – 12.0	2098 – 2219
B₁ (Tsuph = 10 °F)	(49.9, 10.3) – (56.2, 10.3)	27550 – 31017	83.6 – 84.1	8.6 – 9.1	1322 – 1306
B₁ (Tsuph = 15 °F)	(47.2, 15.5) – (54.2, 15.2)	26083 – 29809	83.2 – 83.6	8.6 – 9.5	1330 – 1312
B₁ (Tsuph = 20 °F)	(45.0, 19.9) – (51.9, 19.8)	25092 – 28798	82.9	8.5 – 9.5	1336 – 1308
Eᵥ (Tsuph = 10 °F)	(44.9, 9.8) – (51.4, 9.7)	23999 – 27276	88.6 – 89.2	7.2 – 7.9	1438 – 1418
F₁ (Tsuph = 10 °F)	(45.8, 10.1) – (51.8, 10.2)	27493 – 30737	68.7 – 69.2	8.5	1081 – 1064

[1] – Evaporator exit refrigerant saturation temperature and superheat (T_{evap}, T_{suph})
[2] – Capacity at the temperature conditions listed in column 2 (Col 2)

8.1: Matched condensing unit linear fits

Tables 8.1.1 and 8.1.2 list the linear fits for refrigerant-side capacity and CD unit total power for all the data seen in the following figures. Figures 8.1.1 through 8.1.8 show the effects of varied superheat on the refrigerant-side capacity and CD unit total power as a function of refrigerant saturation temperature at the vapor service valve. Refrigerant-side capacity, in the odd numbered Figures 8.1.1 through 8.1.7, was a weak, but visible, function of superheat over the ranges tested (10 °F to 20 °F). CD unit total power, in the even numbered Figures 8.1.2 through 8.1.8, also showed dependence upon superheat but with less linearity than capacity.

Figures 8.1.9 and 8.1.10 shows refrigerant-side capacity and CD unit power at all standard test conditions at a constant superheat with the points connected by straight lines (these are not linear fits overlayed onto the points).

Table 8.1.1: Linear fits of matched CD unit refrigerant-side capacity as a function of OD service valve vapor refrigerant saturation temperature

Test	Superheat, °F	Number of Points in linear fit	Slope, Btu/(h °F)	Intercept, Btu/h	Pearson's Correlation Coefficient, R^2	Average Subcooling, °F
A_2	10	3	672.97	3184.62	0.99	8.2
A_2	15	5	659.13	4033.72	0.99	8.9
A_2	20	4	663.68	3952.13	0.99	9.8
A_1	10	5	543.10	-2399.960	0.99	6.0
A_1	15	5	535.11	-1720.11	0.99	7.3
A_1	20	5	500.03	222.12	0.99	7.4
B_2	10	3	766.41	1870.97	0.99	10.9
B_2	15	3	746.07	2895.88	0.99	11.3
B_2	20	3	709.55	4791.64	0.99	10.8
B_1	10	3	551.19	-12.458	0.99	8.8
B_1	15	3	535.51	806.41	0.99	9.0
B_1	20	3	535.71	924.27	0.99	9.0
E_v	10	4	511.43	997.40	0.99	7.7
F_1	10	3	535.15	2991.85	0.99	8.4

Table 8.1.2: Linear fits of matched CD power as a function of OD service valve vapor refrigerant saturation temperature

Test	Superheat, °F	Number of Points in linear fit	Slope, W/°F	Intercept, W	Pearson's Correlation Coefficient, R^2	Average Subcooling, °F
A_2	10	3	11.446	1899.41	0.99	8.2
A_2	15	5	11.493	1902.11	0.99	8.9
A_2	20	4	10.679	1947.53	0.99	9.8
A_1	10	5	-2.763	1726.11	0.80	6.0
A_1	15	5	-2.637	1719.74	0.84	7.3
A_1	20	5	-3.549	1767.62	0.99	7.4
B_2	10	3	11.046	1632.55	0.99	10.9
B_2	15	3	11.129	1630.56	0.99	11.3
B_2	20	3	10.747	1649.07	0.99	10.8
B_1	10	3	-2.617	1452.90	0.99	8.8
B_1	15	3	-2.549	1449.37	0.99	9.0
B_1	20	3	-4.056	1520.21	0.97	9.0
E_v	10	4	-3.238	1584.62	0.99	7.7
F_1	10	3	-2.831	1210.27	0.99	8.4

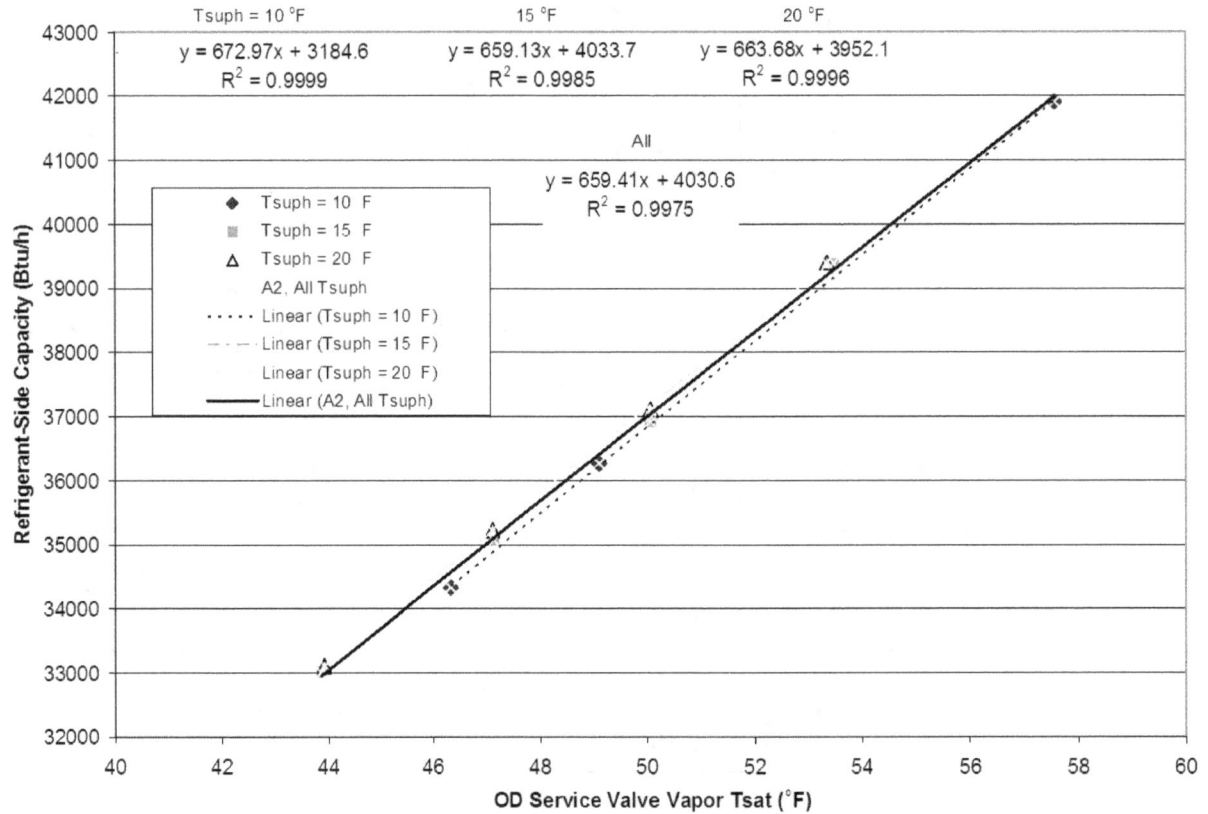

Figure 8.1.1: Matched CD unit A_2 refrigerant-side capacity as a function of OD service valve vapor refrigerant saturation temperature at several superheats

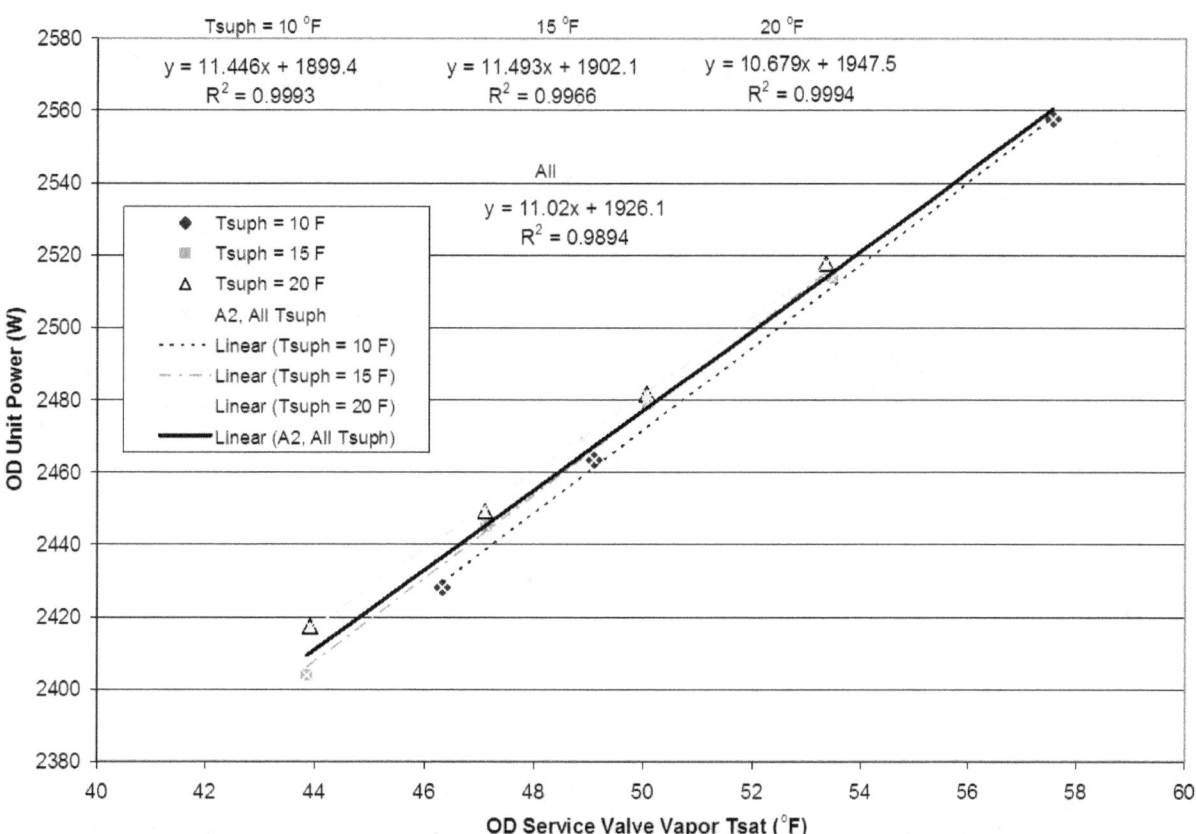

Figure 8.1.2: Matched CD unit A_2 power as a function of OD service valve vapor refrigerant saturation temperature at several superheats

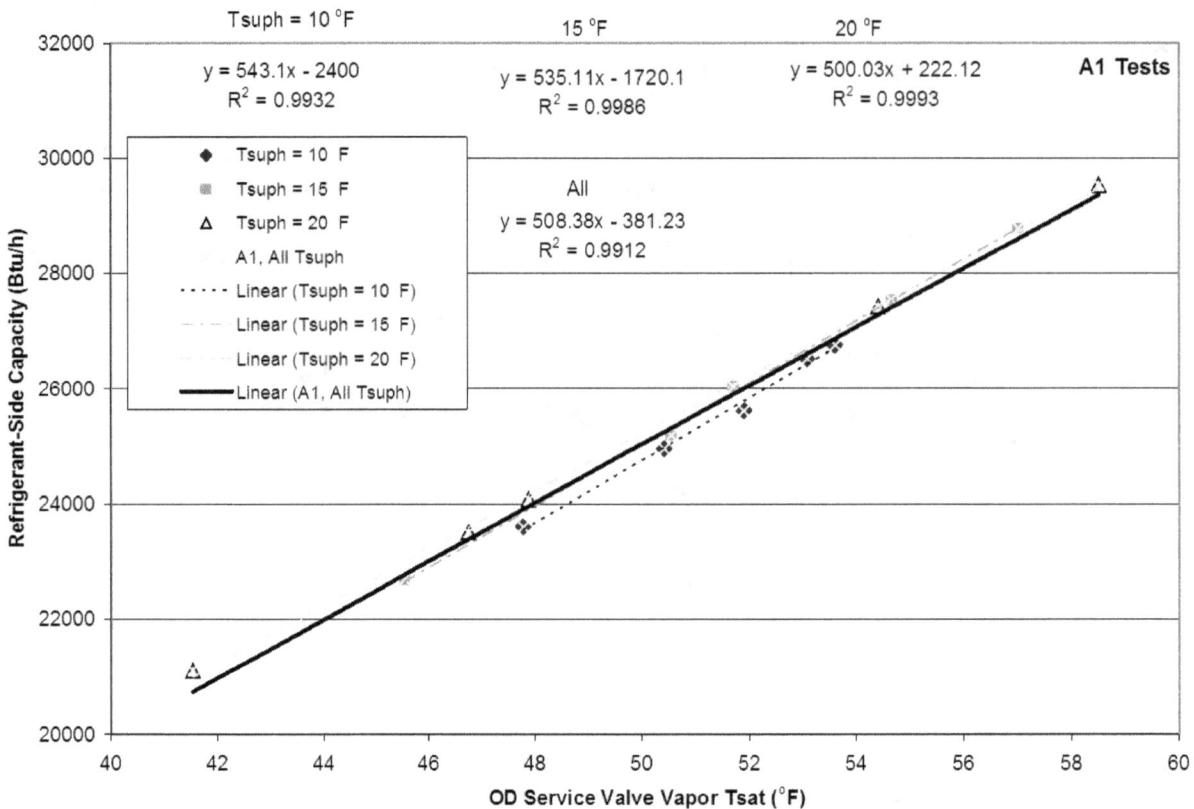

Figure 8.1.3: Matched CD unit A_1 refrigerant-side capacity as a function of OD service valve vapor refrigerant saturation temperature at several superheats

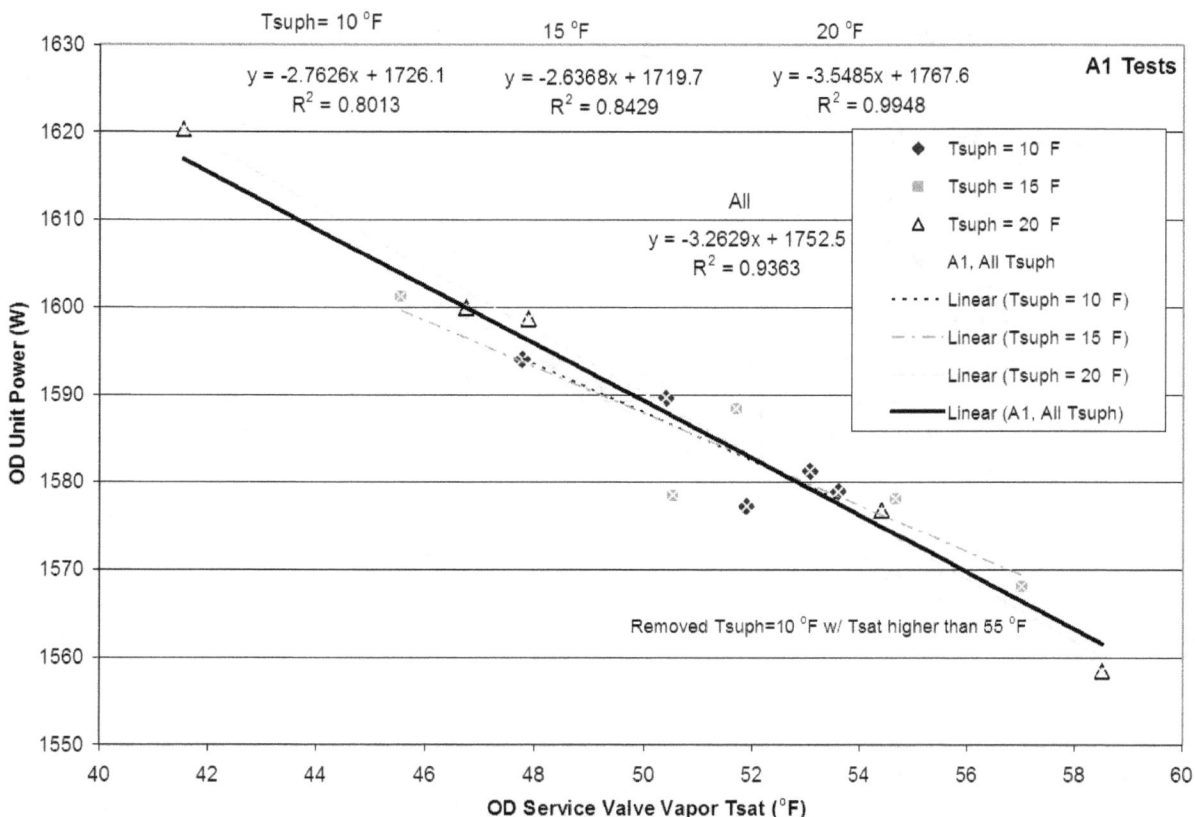

Figure 8.1.4: Matched CD unit A_1 power as a function of OD service valve vapor refrigerant saturation temperature at several superheats

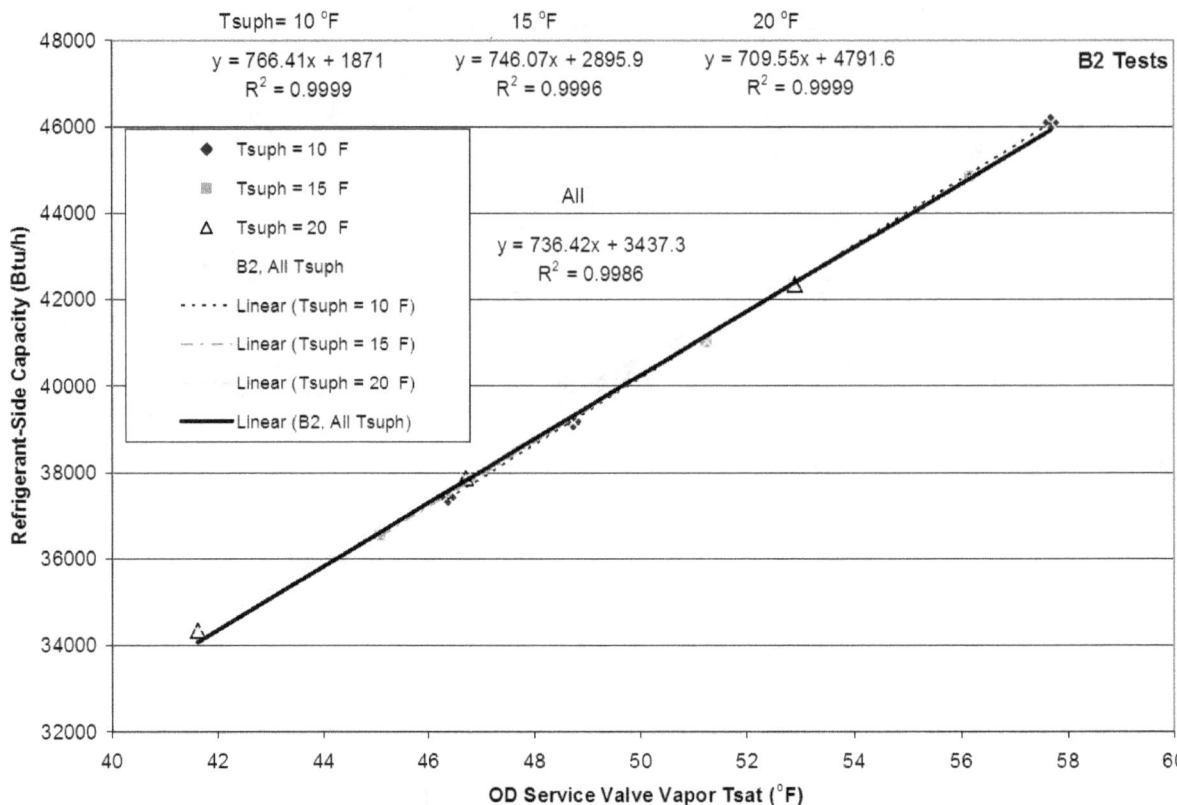

Figure 8.1.5: Matched CD unit B_2 refrigerant-side capacity as a function of OD service valve vapor refrigerant saturation temperature at several superheats

36

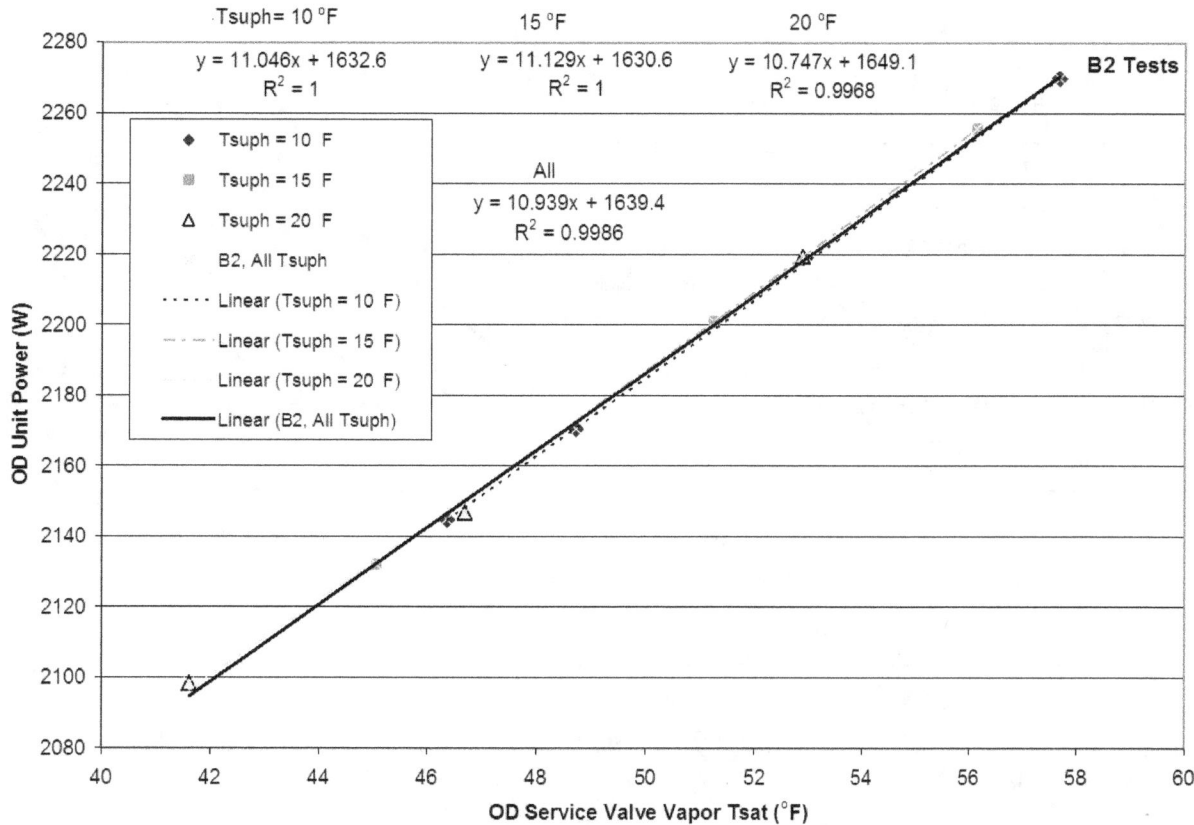

Figure 8.1.6: Matched CD unit B_2 power as a function of OD service valve vapor refrigerant saturation temperature at several superheats

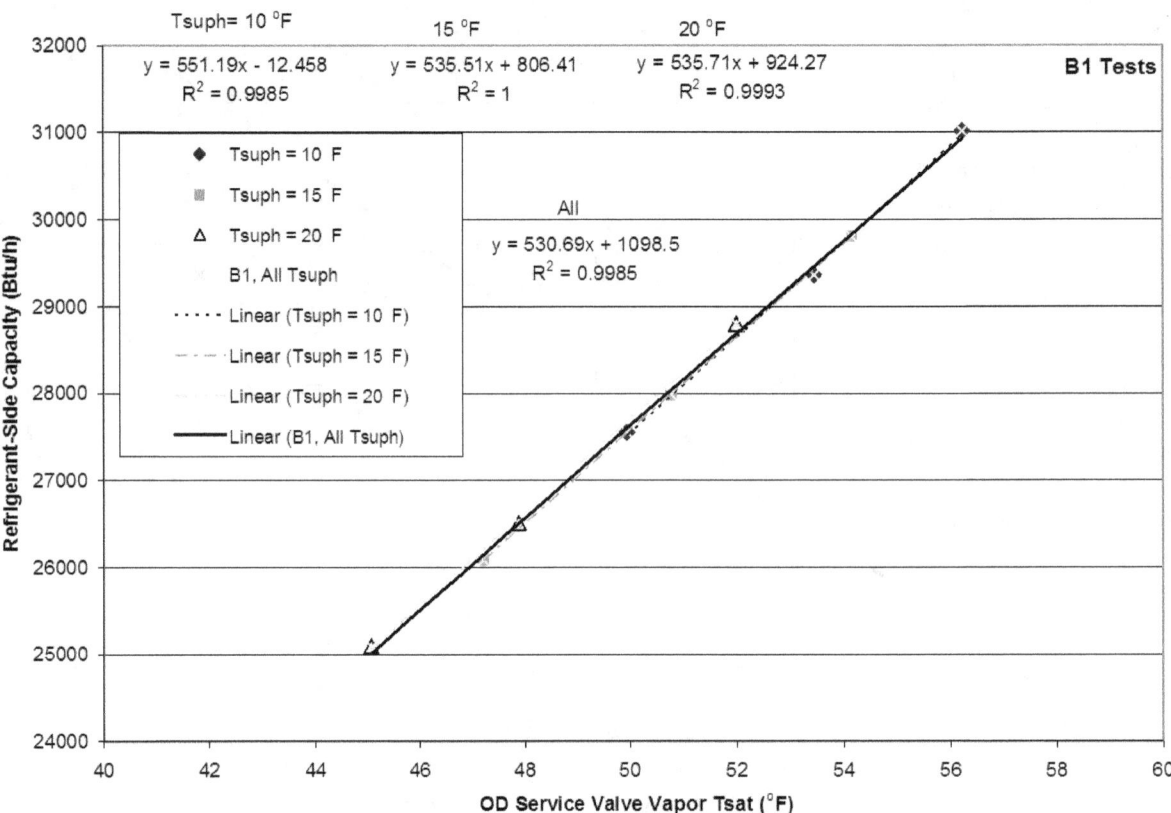

Figure 8.1.7: Matched CD unit B_1 refrigerant-side capacity as a function of OD service valve vapor refrigerant saturation temperature at several superheats

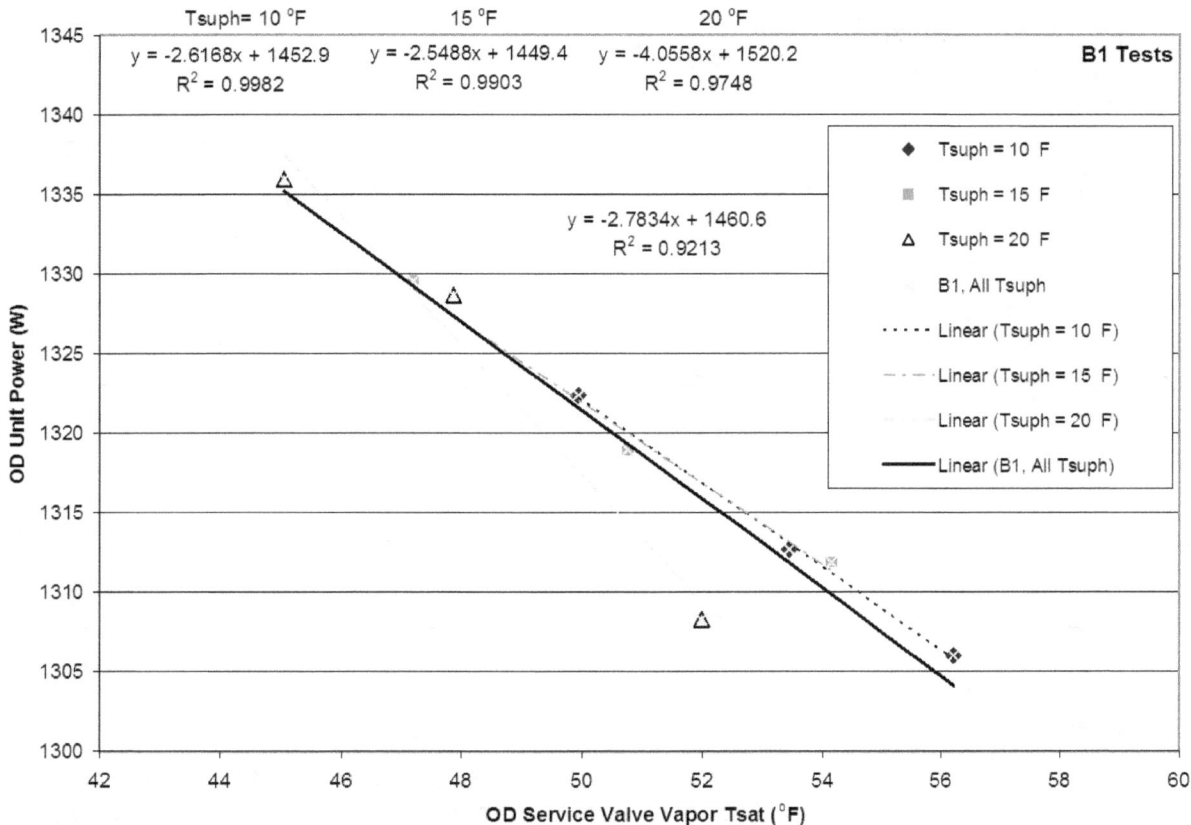

Figure 8.1.8: Matched CD unit B₁ power as a function of OD service valve vapor refrigerant saturation temperature at several superheats

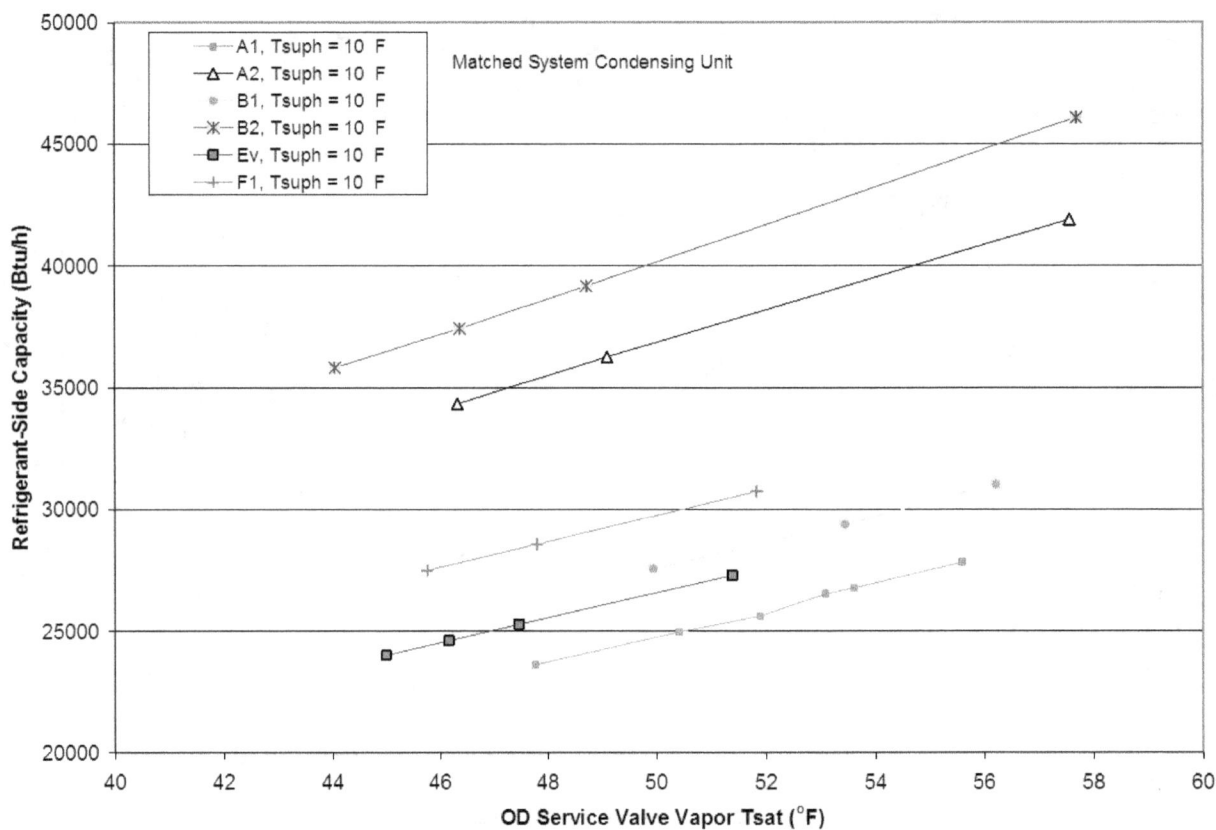

Figure 8.1.9: Matched CD unit capacity for all conditions at a superheat of 10.0 °F

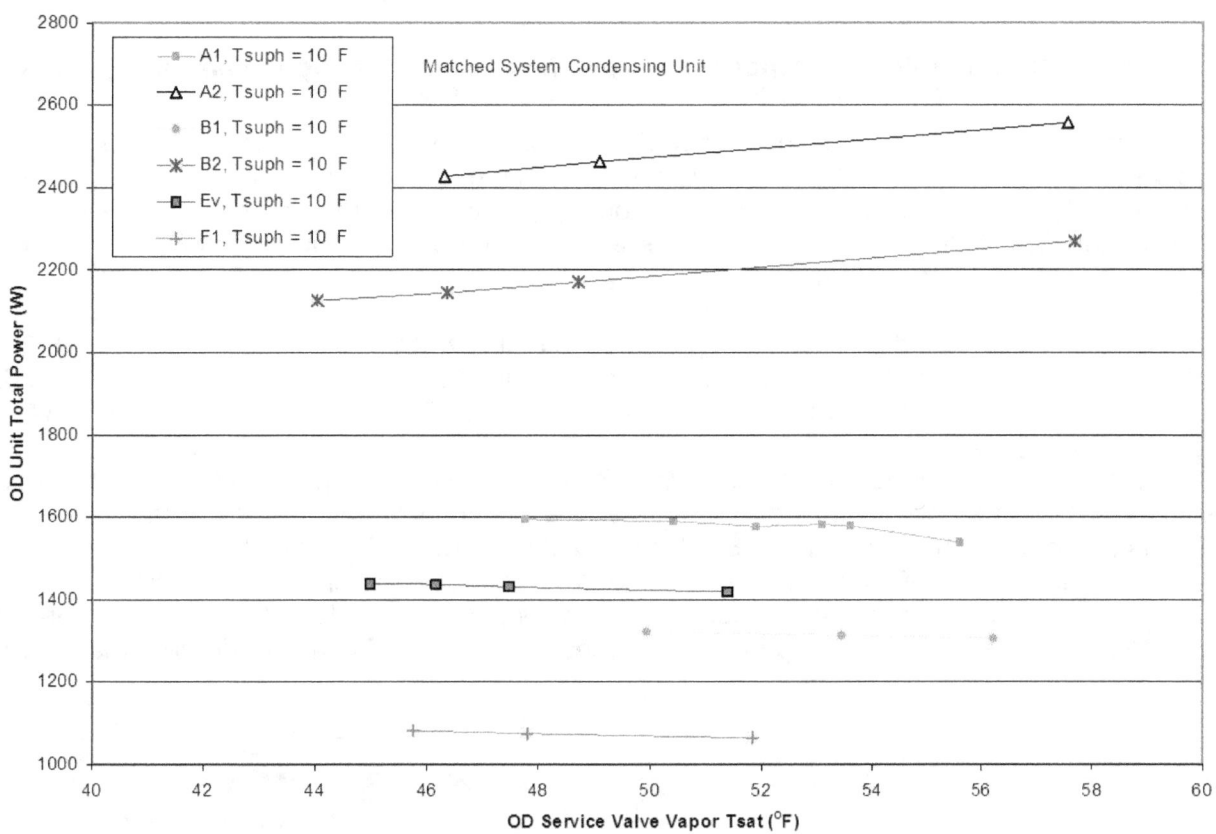

Figure 8.1.10: Matched CD unit power for all conditions at a superheat of 10.0 °F

8.2: Matched CD unit refrigerant mass flow specific capacity (change in enthalpy)

As seen in Figure 8.2.1, condensing unit refrigerant-side capacity is a function of OD service valve vapor refrigerant saturation temperature and superheat, outdoor air temperature and compressor speed. A superheat increase from 10 °F to 20 °F raised specific capacity by approximately 2.2 Btu/lb. Mass flow rate was also affected by the resulting change in suction density seen with the change in superheat.

$$q/mdot = f1(T_{sat}, T_{suph}, T_{od}) \text{ see Figure 8.2.1} \qquad 8.2.1$$
$$mdot = f2(T_{sat}, T_{suph}, n) \text{ see Figure 8.2.3} \qquad 8.2.2$$
$$q = (q/mdot)(mdot) = f1 \cdot f2 = f3(T_{sat}, T_{suph}, T_{od}, n) \qquad 8.2.3$$

Use of a compressor map to predict mass flow rate would allow refrigerant side capacity to be predicted at high and low compressor speeds and associated outdoor airflow rates given the linear fits to the data shown in Figures 8.2.1 and 8.2.2. Figure 8.2.3 shows the refrigerant mass flow rate as a function of evaporator exit refrigerant saturation temperature at high and low airflow rates corresponding to the standard test conditions with two different levels of evaporator exit superheat. The addition of superheat produces a negative offset for refrigerant mass flow rate.

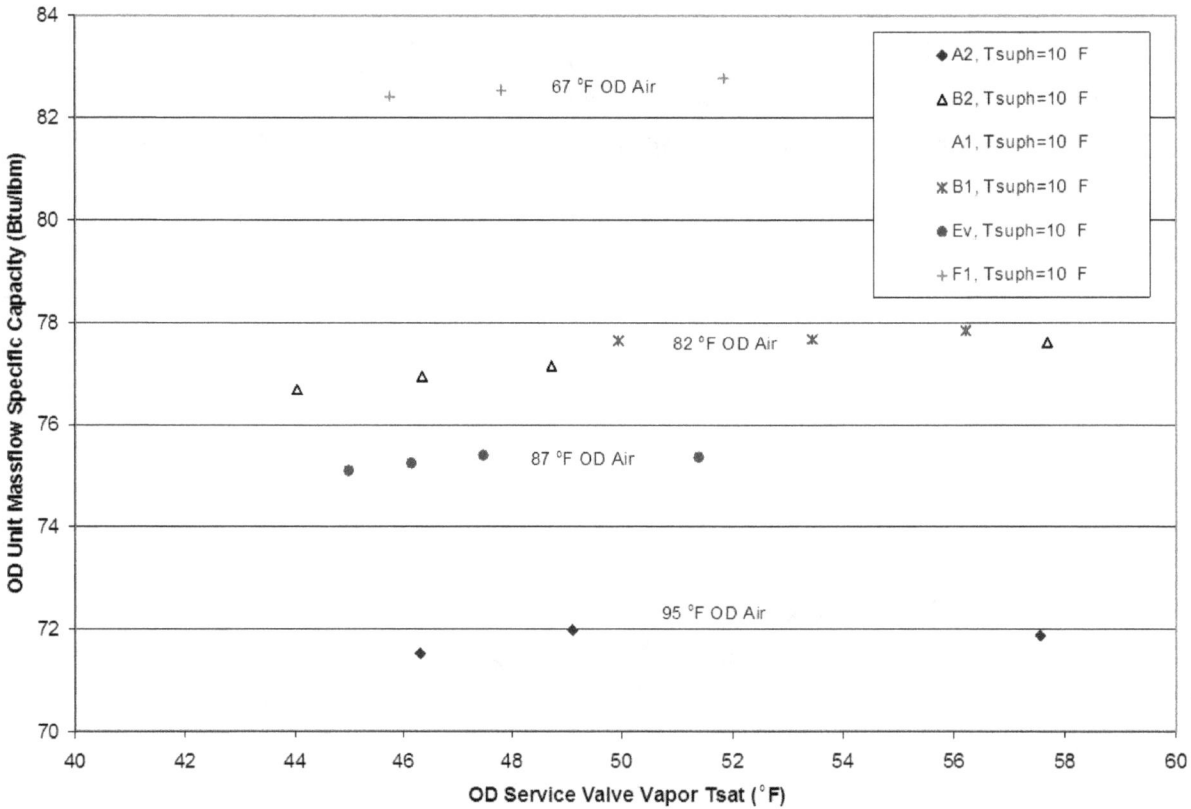

Figure 8.2.1: Matched CD unit capacity per unit of refrigerant mass flow rate at different outdoor air temperatures and constant superheat of 10.0 °F

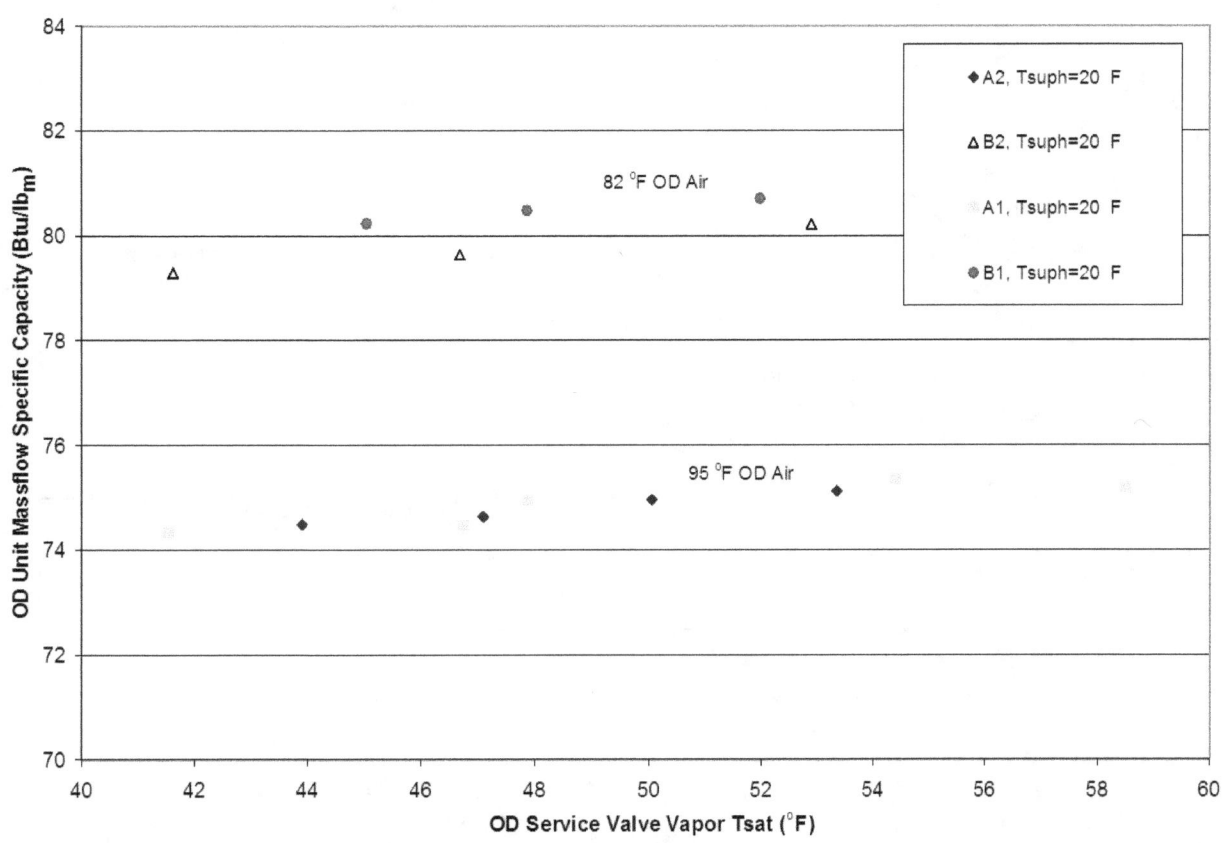

Figure 8.2.2: Matched CD unit capacity per unit of refrigerant mass flow rate at different outdoor air temperatures and constant superheat of 20.0 °F

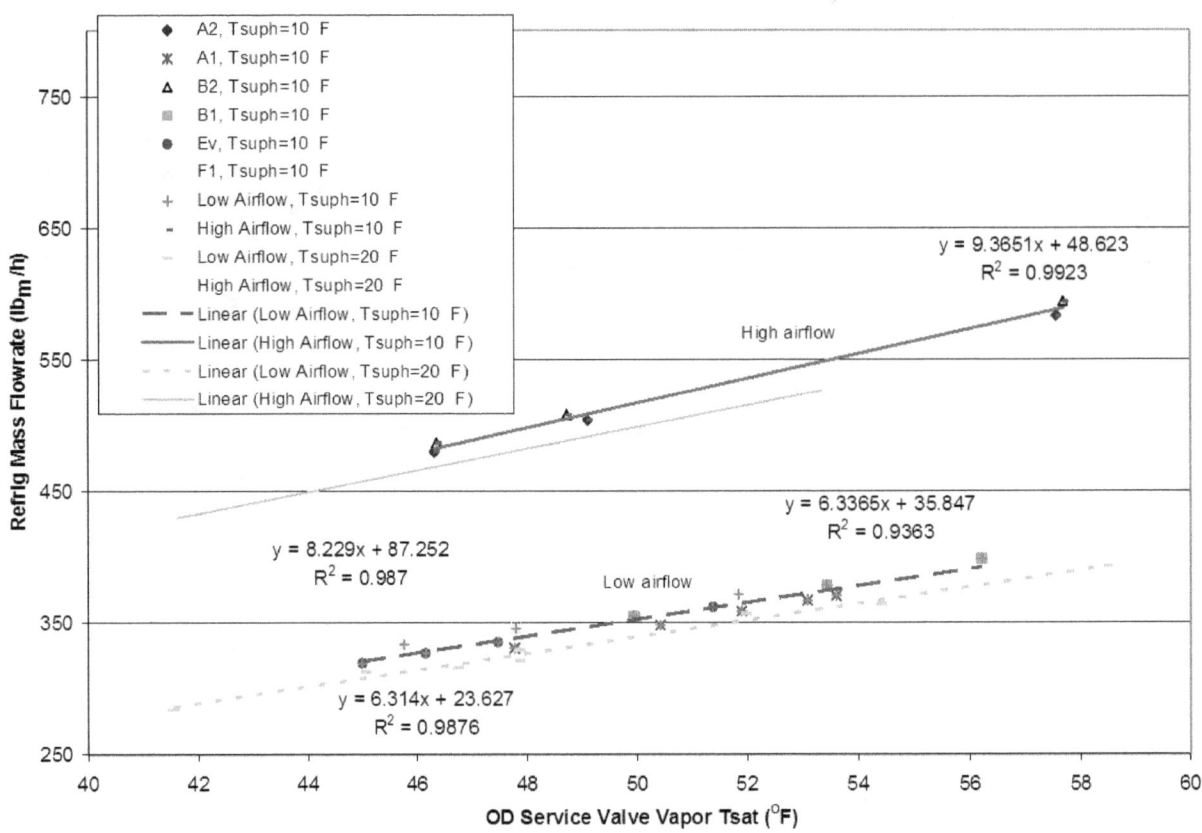

Figure 8.2.3: Matched CD unit refrigerant mass flow rate at high and low compressor speeds

9: COMPARISON OF MEASURED SYSTEM PERFORMANCE TO LINEAR FIT PREDICTIONS

9.1: Calculation of capacity and EER

With the coil capacity coefficients and CD unit capacity coefficients, the linear fit method can be used to calculate cooling capacity and EER for the matched and mixed systems.

The calculation procedure can be implemented computationally by solving the set of two linear equations for the evaporation temperature at which the cooling capacity of the coil equals the cooling capacity of the CD unit:

$$q_{CD} = B_{CD} + A_{CD}T_{evap} = q_{coil} = B_{coil} + T_{evap}A_{coil} \qquad 9.1.1$$

$$T_{evap} = \frac{(B_{coil} - B_{CD})}{(A_{CD} - A_{coil})} \qquad 9.1.2$$

In the equations above, B represents the intercept and A represents the slope for the CD unit (CD subscript) and evaporator coil (coil subscript), respectively. Applying the obtained value of the saturation temperature into either capacity equation yields the capacity of the evaporator. The rated cooling capacity of the system can be obtained by reducing the evaporator capacity by the fan heat. For coils equipped with a fan, the fan heat was measured; for other coils it can be calculated according to AHRI Standard 210/240 (AHRI 2008).

$$Q_{total} = q_{coil} - Q_{ID\ fan} \qquad 9.1.3$$

Similarly, the total power of the system can be obtained by applying the value of the evaporator saturation temperature from Equation 9.1.2 into the condensing unit power Equation 9.1.4 and making adjustment for the indoor fan power as shown in Equation 9.1.5.

$$p_{CD} = b_{CD} + a_{CD}T_{evap} \qquad 9.1.4$$

$$P_{total} = p_{CD} + P_{ID\ fan} \qquad 9.1.5$$

Table 9.1.1 compares the matched and mixed system tests to the linear fit calculated values. Table 9.1.1 uses the linear fits at a constant superheat of 10.0°F. No correction was made for pressure drop in the refrigerant vapor line; no adjustment of evaporator saturation temperature was applied to the CD unit evaporator saturation temperature measured at the service valve.

Liquid refrigerant temperature determines the inlet enthalpy for the evaporator and thus will have some effect upon cooling capacity. This effect was simulated and empirically correlated for R22 and R410A coils in the previous study by Payne and Domanski (2006). In that study, the effects of liquid temperature (and superheat) were included by adjusting the apparent evaporator temperature. In the previous single-speed linear fit method, this empirical correction was applied to adjust the rated cooling capacity, Q(95), and to determine the CD unit power at the corrected T_{evap}. In the case of two-speed and variable speed equipment, the adjustment for liquid temperature and superheat differentce between the mixed coil and matched system condensing unit would be applied for each standard test conditions to correct the T_{evap} for each case. The correction has the following form shown in Equations 9.1.6 through 9.1.10.

Step 1: Estimate the correction for the indoor section capacity equation, ε_{1cor}

$$\varepsilon_{1cor} = \left(\frac{T_{liq,CD}}{T_{liq,coil}}\right)^{-0.123} \left(\frac{T_{suph,CD}}{T_{suph,coil}}\right)^{-0.0879} \qquad 9.1.6$$

where: $T_{liq,CD}$ - refrigerant liquid temperature as listed for the outdoor section at the A Test conditions ($^\circ$F)

$T_{suph,CD}$ - refrigerant superheat at the evaporator exit as listed for the outdoor section at the A Test conditions ($^\circ$F)

$T_{liq,coil}$ - refrigerant liquid temperature used during the generation of the linear fit for the indoor coil ($^\circ$F)

$T_{suph,coil}$ - refrigerant superheat at the evaporator exit used during the generation of the linear fit for the indoor coil ($^\circ$F)

Step 2: Estimate the evaporator refrigerant saturation temperature at the standard test conditions, T_{evap}

$$T_{evap} = \frac{\varepsilon_{1cor} \cdot C_{coil} - C_{CD}}{D_{CD} - \varepsilon_{1cor} \cdot D_{coil}} \qquad 9.1.7$$

Step 3: Improve the estimate of the correction for indoor section capacity equation, ε_{2cor}

$$\varepsilon_{2cor} = \left(\frac{T_{liq,CD}}{T_{liq,coil}}\right)^{b1} \left(\frac{T_{suph,CD}}{T_{suph,coil}}\right)^{b2} \qquad 9.1.8$$

where: $b1 = -0.123\left(\frac{T_{evap}}{50}\right)$

$b2 = -0.0879\left(\frac{T_{evap}}{50}\right)$

$T_{evap}()$ - evaporator refrigerant saturation temperature calculated from Equation 9.1.7, converted to $^\circ$F (if calculated in $^\circ$C)

Step 4: Calculate evaporator refrigerant saturation temperature at the standard test conditions, T_{evap}.

$$T_{evap}(\;) = \frac{\varepsilon_{2cor} \cdot C_{coil} - C_{CD}}{D_{CD} - \varepsilon_{2cor} \cdot D_{coil}} \qquad 9.1.9$$

Step 5: Calculate mixed system capacity at the standard test conditions, Q_{mixed}.

$$Q_{mixed} = q_{CD} - Q_{fan,mixed} = C_{CD} + D_{CD} \cdot T_{evap} - Q_{fan,mixed} \qquad 9.1.10$$

Table 9.1.2 applies the correction to some mixed coil linear fits determined at different superheats. For the tests shown, the matched CD unit linear fit at 10 $^\circ$F is used and the corrected evaporator saturation temperature is calculated.

46

Table 9.1.1: System capacities, total power, and EER from the linear fit method (all systems included an indoor blower)

Type	T_{evap}, °F	Meas. T_{evap}, (°F)	Indoor Airflow, (scfm)	P_{fan}, W	Q_{fan}, (Btu/h) (1)(2)	p_{CD}, W	Meas. p_{CD}, W	q, (Btu/h)	Meas. q, (Btu/h)	Q, (Btu/h)	Meas. Q, (Btu/h)	EER, Btu/(Wh)	Meas. EER, Btu/(Wh)	p_{CD} % error	Q % error	EER % error
Matched																
A_2	49.90	52.98	1241	279	951.3	2471	2497	36766	37404	35815	36453	13.027	13.130	-1.1	-1.7	-0.8
A_2	49.90	53.01	1242	284	969.8	2471	2508	36766	37131	35796	36161	12.994	12.949	-1.5	-1.0	0.4
A_2	49.90	52.03	1242	277	945.1	2471	2509	36766	36573	35821	35628	13.037	12.787	-1.5	0.5	2.0
A_1	51.22	54.10	942	112	382.8	1585	1593	25419	25689	25036	25307	14.755	14.845	-0.5	-1.1	-0.6
B_1	50.29	53.31	943	113	384.4	1321	1321	27709	27581	27324	27196	19.055	18.972	0.0	0.5	0.4
B_1	50.29	52.39	906	65	221.1	1321	1325	27709	27992	27488	27771	19.831	19.980	-0.3	-1.0	-0.7
B_2	48.81	51.59	1234	277	946.4	2172	2206	39279	39938	38333	38991	15.652	15.701	-1.6	-1.7	-0.3
B_2	48.81	51.14	1244	293	1000.4	2172	2171	39279	39803	38279	38802	15.529	15.749	0.0	-1.3	-1.4
Mixed #1																
B_2	44.60	46.38	1222	392	1338.3	2125	2136	36053	36219	34715	34880	13.790	13.794	-0.5	-0.5	0.0
B_2	44.60	46.49	1223	396	1352.4	2125	2140	36053	36282	34700	34929	13.761	13.772	-0.7	-0.7	-0.1
B_1	46.80	48.43	960	237	808.5	1330	1332	25784	25138	24976	24330	15.934	15.505	-0.1	2.7	2.8
B_1	46.80	48.56	976	249	850.3	1330	1326	25784	25122	24934	24272	15.784	15.407	0.3	2.7	2.4
B_1	46.80	48.47	976	251	856.9	1330	1322	25784	25175	24927	24318	15.761	15.457	0.6	2.5	2.0
A_2	45.24	47.31	1209	391	1334.7	2417	2442	33632	33404	32297	32069	11.500	11.320	-1.0	0.7	1.6
A_2	45.24	46.96	1216	384	1311.0	2417	2455	33632	33177	32321	31865	11.537	11.223	-1.5	1.4	2.8
A_1	48.01	49.13	963	242	824.2	1593	1599	23673	23235	22849	22411	12.451	12.173	-0.4	2.0	2.3
A_1	48.01	49.32	964	244	832.2	1593	1603	23673	23125	22841	22293	12.431	12.072	-0.6	2.5	3.0
A_1	48.01	48.84	969	236	804.8	1593	1601	23673	23002	22868	22197	12.501	12.083	-0.5	3.0	3.5
Mixed #2																
A_1	48.62	51.38	752	584	1993.8	1592	1591	24005	24297	22011	22303	10.115	10.255	0.1	-1.3	-1.4
A_1	48.62	51.42	753	586	1997.8	1592	1593	24005	24222	22007	22224	10.107	10.203	-0.1	-1.0	-0.9
A_2	45.5	46.31	750	587	2002.6	2420	2451	33777	31984	31774	29981	10.568	9.868	-1.3	6.0	7.1
B_1	47.3	50.07	753	586	1998.6	1329	1318	26048	25862	24049	23864	12.559	12.534	0.8	0.8	0.2
B_2	44.6	45.25	760	593	2024.4	2125	2129	36033	34677	34009	32652	12.511	11.993	-0.2	4.2	4.3

47

Table 9.1.2: Condensing unit capacity without and with corrected T_{evap}

Test	Uncorrected $T_{evap.}$	T_{liq} coil	T_{liq} CD unit	T_{suph} coil	T_{suph} CD unit	ε_{1cor}	First correction $T_{evap.}$	ε_{2cor}	Final corrected $T_{evap.}$	Uncorrected q_{CD}	Corrected q_{CD}	% change
A_2, matched	47.71	94.9	97.1	20.4	10.0	1.0617	48.3348	1.0622	48.34	35290.5	35715.9	1.2
A_2, matched	51.09	95.1	97.1	5.2	10.0	0.9417	50.5264	0.9433	50.54	37565.1	37198.5	-1.0
B_2, matched	46.39	82.1	84.5	20.2	10.2	1.0581	47.1095	1.0579	47.11	37422.1	37973.8	1.5
A2, mixed #2	45.64	99.9	97.1	10.1	10.0	1.0044	45.7017	1.0011	45.66	33898.5	33909.4	0.03
A2, mixed #2	45.69	105.1	97.1	10.2	10.0	1.0115	45.8034	1.0025	45.67	33932.9	33922.5	-0.03
F1, mixed #2	45.98	73.1	70.5	10.1	10.2	1.0036	46.0344	0.9996	45.98	27600.3	27597.4	-0.01

48

9.2: Calculation of SEER

Linear fit method SEER may be directly calculated using the bin method if the matched system CD unit linear fits for capacity and power are provided. The rater only needs linear fits for the mixed coil capacity at the standard conditions with corresponding indoor blower power. Table 9.2.1 shows the calculated SEER for the matched system and two mixed systems using the linear fits at a superheat of 10 °F and providing the indoor fan power correction for the matched system (Figure 9.2.1).

Figure 9.2.2 shows the effect of varying the cyclic degradation coefficient from a value of 0.05 to 0.25; there is a range of approximately 7.5 % with respect to the SEER values at a C_D=0.25.

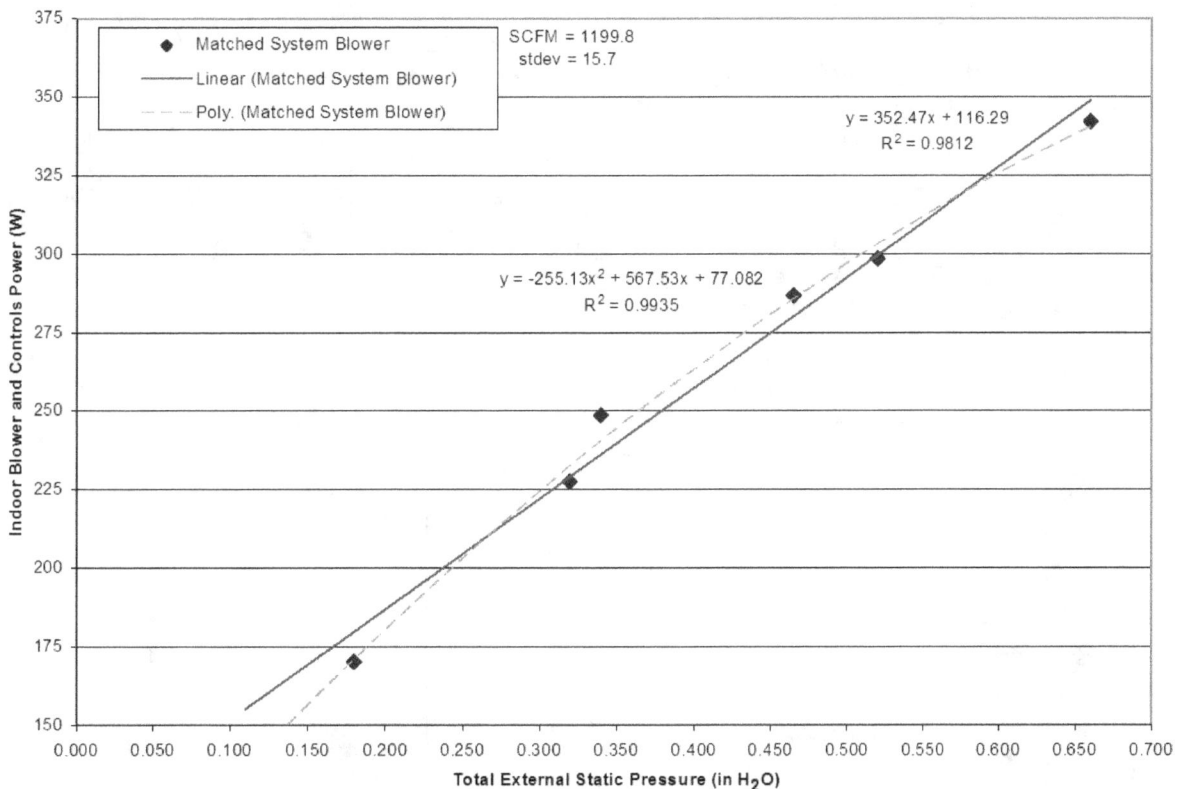

Figure 9.2.1: Matched system, high speed, blower power as a function of external static pressure at constant airflow rate

Table 9.2.1: Bin method SEER calculated using linear fits at 10 °F superheat

Type	T_{evap}, °F	Indoor Airflow, scfm	P_{fan}, W	Q_{fan}, Btu/h	p_{CD}, W	q, Btu/h	Q, Btu/h	Total Power, W	EER, Btu/Wh	SEER, Btu/Wh
Matched w/ C_D=0.25										
A_2	49.9	1241	170	580	2471	36766	36186	2641	13.70	
B_2	48.8	1234	170	580	2172	39279	38699	2342	16.53	17.64
B_1	50.3	943	70	239	1321	27709	27470	1391	19.74	
F_1	50.0	942	70	239	1069	29754	29515	1139	25.92	
Mixed #1 w/ C_D=0.25										
A_2	45.2	1209	391 [2]	1334.1	2417	33632	32298	2808	11.50	
B_2	44.6	1222	392	1337.5	2125	36053	34715	2517	13.79	14.26
B_1	46.8	960	237	808.6	1330	25784	24976	1567	15.93	
F_1	45.9	960	237	808.6	1080	27566	26758	1317	20.31	
Mixed #2 w/ C_D=0.25										
A_2	45.5	750	587	2002.8	2420	33777	31774	3007	10.57	
B_2	44.6	760	593	2023.3	2125	36033	34010	2718	12.51	11.43
B_1	47.3	753	586	1999.4	1329	26048	24048	1915	12.56	
F_1	45.9	753	586	1999.4	1080	27557	25558	1666	15.34	

[1]- All data taken at 208 VAC, single-phase power.
[2]- No fan power credit given for Mixed #1 system.

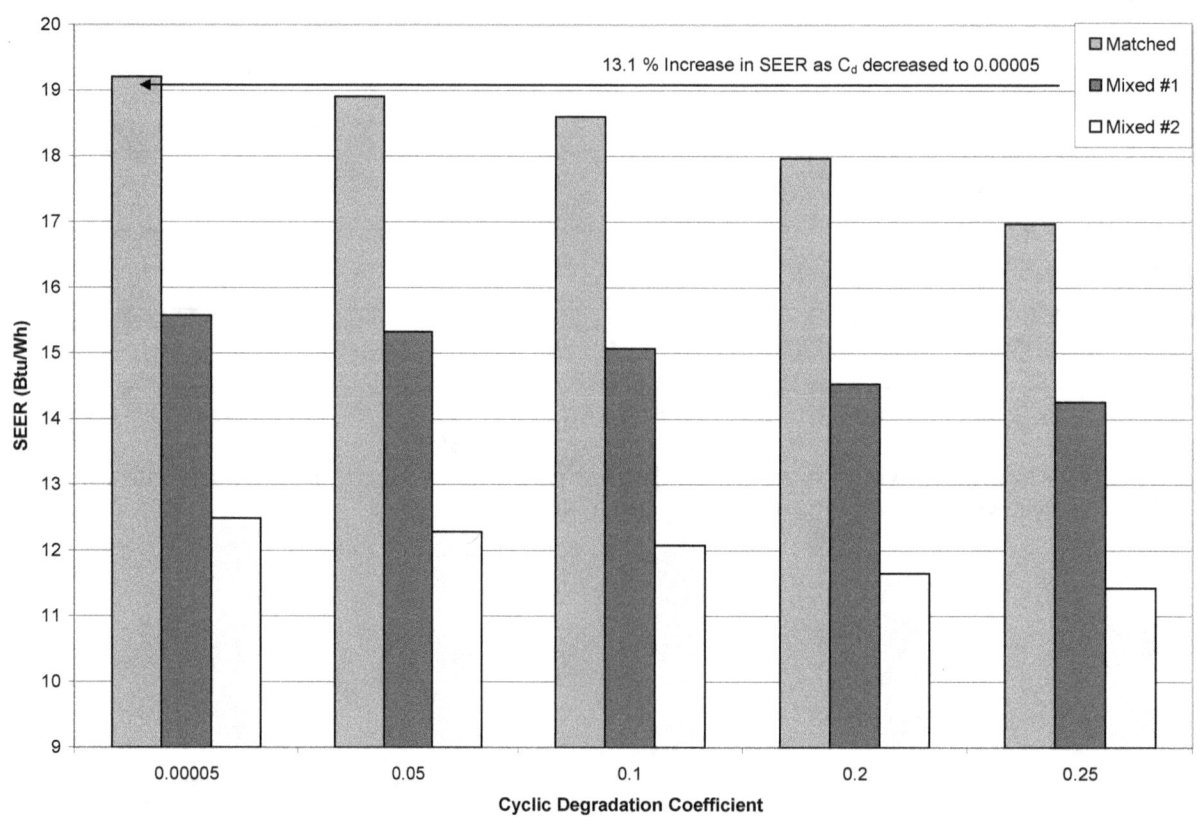

Figure 9.2.2: SEER for the two-speed systems calculated with varied cyclic degradation coefficients

Table 9.2.2 shows the results of using Equation 1.5 (shown below) to scale the matched system SEER to determine the mixed systems' SEER. F_{exp} was determined using Tables 9.2.3 and 9.2.4 as shown in the single-speed linear fit method (Payne and Domanski 2006). The manufacturer's rated SEER for the matched system is 20. No cyclic testing was done to determine the actual degradation coefficient, and testing voltage was 208 VAC. Once the fan power credit is given to the matched system, and a low Cd value is used, the bin SEER becomes 19.203 even at the lower voltage test conditions. This is within 5 % of the 20 SEER claimed for the matched system. Mixed system #2 would have had a similar increase in SEER if the fan power correction was performed, but the fan power curve at constant CFM was not produced. The key point to note through all of this testing is that the performance of the matched and mixed systems is very linear; this means the linear fit method of scaling the matched system SEER to calculate the mixed system SEER will work.

$$SEER_{mixed} = SEER_{matched} \frac{\sum EER_{j,\,mixed}}{\sum EER_{j,\,matched}} F_{exp} \qquad \text{copy of 1.5}$$

Table 9.2.2: Scaled SEER calculated using linear fits at 10 °F superheat with $C_d = 0.00005$ and $C_d = 0.25$

Type	EER, Btu/Wh	$\sum EER_{mixed}$	$\sum EER_{matched}$	F_{exp}	$\frac{\sum EER_{j,\,mixed}}{\sum EER_{j,\,matched}} F_{exp}$	Scaled SEER, Btu/Wh	Bin SEER, Btu/Wh	% diff wrt Bin Method
A_2	13.704	75.894 Matched	75.894	1.0	100 %	19.203	19.203	0.0 %
B_2	16.526							
B_1	19.744					16.973	16.973	0.0 %
F_1	25.920							
A_2	11.501	61.539 Mixed #1	75.894	1.0	81.1 %	15.571	15.577	-0.04 %
B_2	13.791							
B_1	15.934					13.763	14.26	-3.49 %
F_1	20.313							
A_2	10.568	50.976 Mixed #2	75.894	1.0	67.2 %	12.898	12.285	3.30 %
B_2	12.513							
B_1	12.557					11.400	11.431	-0.27 %
F_1	15.338							

Dougherty (2003), working with DOE and AHRI, performed a statistical analysis of experimentally determined C_D values for a large sample of systems. He grouped the studied systems into four basic categories shown in Table 9.2.3. The analysis of C_D values for these four system categories produced the C_D percentiles shown in Table 9.2.4. Using the 95th percentile values for each system category in Table 9.2.4, in addition to Domanski's (1989) empirical correction for time delay relays and different expansion devices, yields the F_{exp} values in Table 9.2.5.

Table 9.2.3: System classifications for cyclic degradation coefficient analysis (Dougherty 2003)

System Category	Equalize During Off Cycle	Indoor Fan Off Delay	System Components
A	Yes	No	Cap Tube Orifice Bleed TXV
B1	No	No	Non-Bleed TXV Electronic Expansion Device Liquid Line Solenoid
B2	Yes	Yes	Cap Tube Orifice Bleed TXV
C	No	Yes	Non-Bleed TXV Electronic Expansion Device Liquid Line Solenoid

Table 9.2.4: Categorized cyclic degradation coefficient values (Dougherty 2003)

Percentile	A	B1	B2	C
99th	0.24	0.16	0.22	0.15
95th	0.22	0.14	0.14	0.12
90th	0.16	0.14	0.12	0.10
85th	0.14	0.12	0.11	0.09
80th	0.12	0.12	0.10	0.08
75th	0.12	0.11	0.10	0.07
70th	0.11	0.11	0.09	0.06
60th	0.10	0.9	0.08	0.05
50th	0.09	0.07	0.07	0.04
Sample Size	77	58	109	78

Table 9.2.5: F_{exp} for various mixed and matched system combinations

		Matched System			
		A	B1	B2	C
Mixed System	A	1.000	0.990	0.990	0.974
	B1	1.010	1.000	1.000	0.985
	B2	1.010	1.000	1.000	0.985
	C	1.026	1.016	1.016	1.000

10: A DISCUSSION ON BLOWER EFFICIENCY

An interesting analysis of fan power and efficiency was performed by Messmer (2010). His analysis examines the AHRI 210/240 default for fan power per unit airflow rate (0.365 W/scfm) and how this can be related to fan static pressure rise, fan mechanical efficiency, and fan motor efficiency. In his analysis he illustrates the some of the possible assumptions about the fan performance that lead to a value of 0.365 W/scfm. He showed that the following three assumptions produced the default fan power of 0.365 W/scfm:

1) Blower efficiency = 0.55
2) Motor efficiency = 0.55
3) External static pressure across blower = 0.94 inches of water.

Messmer pointed out that high efficiency products, with variable-speed blower motors, will easily require lower than the default fan power. One reason for this is the higher motor efficiency (0.75 to 0.85) and "flat" nature of the fan curves resulting in a low speed mechanical efficiency of approximately 0.50. Thus at low speed, Messmer's calculations showed a fan power per unit airflow rate of approximately 0.157 W/scfm.

For the testing performed at NIST the matched system and mixed system #1 used a variable speed, high efficiency blower motor, while mixed system #2 used a single-speed blower found in small duct, high velocity systems. Table 10.1 shows high and low speed fan efficiency for the airflow rates and fan powers seen in Table 9.2.1. As noted by Messmer, the high efficiency blower in the matched system produced fan power per unit of airflow rate very close to his calculations.

Table 10.1: Fan efficiency for matched and mixed air handlers

Type	Indoor Airflow, scfm	$P_{fan,}$ W	External Static Pressure, in H$_2$O	Blower, W/scfm	Blower, scfm/W
Matched					
A$_2$	1241	279	0.24	0.22	4.45
B$_2$	1234	277	0.24	0.22	4.45
B$_1$	943	113	0.24	0.12	8.35
F$_1$	942	112	0.24	0.12	8.41
Mixed #1					
A$_2$	1209	391	0.2	0.32	3.09
B$_2$	1222	392	0.2	0.32	3.12
B$_1$	960	237	0.21	0.25	4.05
F$_1$	960	237	0.21	0.25	4.05
Mixed #2					
A$_2$	750	587	2.2	0.78	1.28
B$_2$	760	593	2.3	0.78	1.28
B$_1$	753	586	2.2	0.78	1.28
F$_1$	753	586	2.2	0.78	1.28

In another attempt to look at the performance of the electronically commutated blower motor in the matched system, fan power was recorded for constant airflow rate at various external static pressures. The resulting plot was shown in Figure 9.2.2 with power as the ordinate and in Figure 10.1 with power per unit airflow rate as the ordinate. The linear fit correlation coefficients show that this blower setup is very linear over this range of external static pressures. Using the linear fit of Figure 10.1, the external static pressure at 0.365 W/scfm would equal 0.93 inches of water gage. This result is as predicted by Messmer's analysis summarized above.

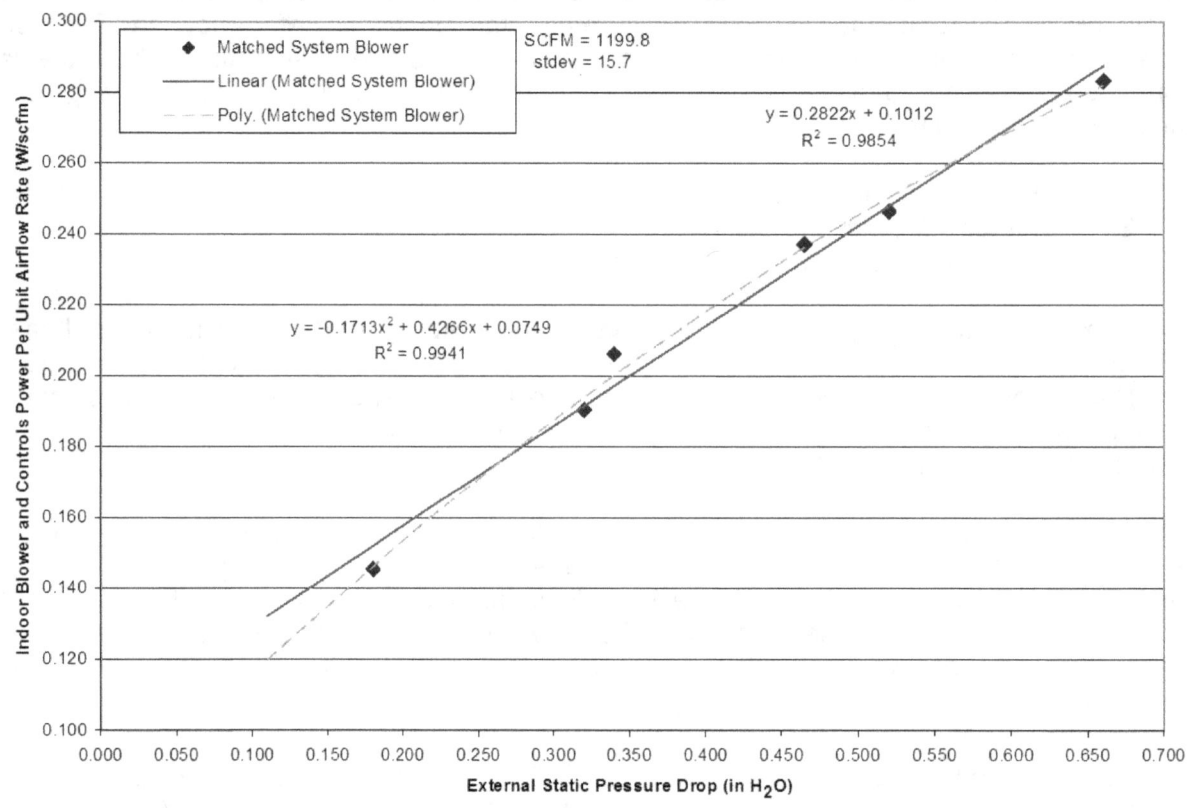

Figure 10.1: Matched system, high speed, blower power per unit of airflow rate as a function of external static pressure at constant airflow rate

11: CONCLUDING REMARKS

Coil cooling capacity was examined for the matched system coil and two mixed system coils. Liquid refrigerant temperature was varied and shown to have a weak effect on linear fit slopes, but this effect may not be negligible. The previous correction method developed by Payne and Domanski (2006) was used to correct the calculated mixed system evaporator temperature for several example tests. For those coil linear fits determined at liquid refrigerant temperatures and superheats different from the matched system CD unit, the corrections moved capacity in the right direction and corresponded to past trends seen with mixed system testing. Superheat correction was applied for several examples and shown to be in the correct direction, but the magnitude of this correction may not be sufficient in all cases. Coil manufacturers would need to modify this superheat correction to produce better agreement in cases where superheat was substantially different (more than 5 °F) from that used to generate the matched CD unit linear fits.

A thorough examination of subcooling should be performed for the CD unit. For the tests presented here, the CD unit charge was set at the A_2 test conditions and then remained unchanged. Subcooling will affect compressor power and thus EER, but different subcoolings were not investigated here.

An attempt was made to normalize the coil's cooling capacity by examining the ratio of coil capacity to standard airflow rate (Btu/(h scfm)). The airflow specific capacity trends, at a given superheat for the matched and mixed coils, were extremely linear even when all liquid temperatures were included in the figures. This type of coil capacity normalization may be useful to determine whether a certain coil is being rated consistently as it is applied to different manufacturer's CD units. A linear fit of airflow specific capacity at a fixed superheat may be generated from two points (possibly taken from the AHRI database). If a particular mixed system utilizing this coil produces airflow specific capacity that is outside an acceptable limit, then the mixed system ratings may be suspect.

A similar attempt was made to normalize condensing unit cooling capacity by dividing the refrigerant-side capacity by the refrigerant mass flow rate. This quantity equals the change in enthalpy of the refrigerant as it passes through the CD unit [(Btu/h) / (lb/h) = Btu/lb]. Condensing unit data showed that the mass flow rate specific capacity was approximately constant at a fixed superheat for a given outdoor air temperature regardless of compressor speed. Therefore, someone could easily use a compressor map to determine refrigerant mass flow rate at a specific evaporator saturation temperature and apply this mass flow rate to the specific capacity line at the appropriate outdoor air conditions to determine refrigerant-side capacity. This kind of analysis was presented only as a possible means for CD unit manufacturers to characterize their product's performance.

A cursory examination of default fan power was presented to illustrate the assumptions necessary to produce the default 0.365 W/scfm mandated by the AHRI 210/240 test procedure. The analysis presented by Messmer (2010) was confirmed by blower power measurements done during this testing; high efficiency, variable-speed, ECM blower performance may be predicted with knowledge of the motor efficiency, blower wheel mechanical efficiency, and static pressure drop. Testing of the matched system blower at constant airflow rate showed that power was very linear over a wide range of external static pressures. The measurements also indicated that the default of 0.365 W/scfm greatly exceeded the matched systems 0.143 W/scfm (linearly extrapolated down to 0.15 in H2O). This gross overestimate of fan power by the default value necessitates that ICM's purchase the variable-speed air handlers with which they want to rate their coil in order to determine reasonably accurate SEER values for the mixed system. This seems to be overly burdensome especially when a more reasonable default power could be calculated.

In lieu of performing a calculation for the default fan power, a statistical analysis of the various adjustable speed blowers found in the AHRI database could be performed. This type of analysis would be similar to that performed by Dougherty (2003) when he presented an analysis of the cooling mode, cyclic degradation coefficients determined in the AHRI test program; blower efficiency (W/scfm) could be statistically analyzed and grouped by relevant blower parameters and characteristics. Such an analysis could incorporate adjustable speed air handlers and furnace blowers. The ease of acquiring detailed fan power data is questionable, but this type of analysis would be foolproof in that it would only look at measured results and not attempt a calculation of default fan power.

The results of this investigation will be used to produce a detailed test procedure similar to Payne and Domanski (2006). This type of procedure is meant to guide raters in developing a linear fit based Alternate Rating Method (ARM) for their mixed systems. The rater is free to modify and use parts of any procedure to create an ARM that specifically applies to their products.

REFERENCES

ASHRAE 1988. ANSI/ASHRAE Standard 37. *Methods of testing for rating unitary air conditioning and heat pump equipment*. American Society of Heating, Refrigerating and Air-Conditioning Engineers. 1791 Tullie Circle NE, Atlanta, GA, USA.

AHRI 2008. Standard 210/240, *Standard for unitary air-conditioning and air-source heat pump equipment*, Air-Conditioning, Heating, and Refrigeration Institute, 4100 North Fairfax Drive, Suite 200, Arlington, VA 22203.

CFR, 2009a. *Code of Federal Regulations*, Part 430, Appendix M to Subpart B, Uniform test method for measuring the energy consumption of central air conditioners, Office of the Federal Register, National Archives and Records Administration, Washington, DC. http://www.gpoaccess.gov/cfr/retrieve.html

CFR, 2009b. *Code of Federal Regulations*, Part 430.24, Subsection m1, Units to be tested, Federal Register, National Archives and Records Administration, Washington, DC.

Dougherty, B., 2003. *New defaults for the cyclic degradation coefficient used in rating air conditioners and heat pumps*, Seminar 40, ASHRAE Annual Meeting in Kansas City, KS, July 2.

Domanski, P.A., 1989. *Rating procedure for mixed air-source unitary air conditioners and heat pumps operating in the cooling mode – revision 1*, NISTIR 89-4071, U. S. Dept of Commerce, Natn'l Institute of Standards and Technology, Gaithersburg, Maryland USA 20899.

Messmer, C.S., 2010. Personal communications. Unico System, 326 Bluff View Circle, Saint Louis, MO 63129-5060, USA.

Payne, W.V. and Domanski, P.A., 2005. *A curve-based mixed system rating method for unitary air conditioners*, NISTIR 7225, U. S. Dept of Commerce, Natn'l Institute of Standards and Technology, Gaithersburg, Maryland USA 20899.
http://www.bfrl.nist.gov/863/HVAC/pubs/index.htm

Payne, W.V. and Domanski, P.A., 2006. *Linear fit-based rating procedure for mixed air-source unitary air conditioners and heat pumps operating in the cooling mode*, NISTIR 7325, U. S. Dept of Commerce, Natn'l Institute of Standards and Technology, Gaithersburg, Maryland USA 20899.
http://www.bfrl.nist.gov/863/HVAC/pubs/index.htm

Taylor, B. N., and Kuyatt, C. E., 1994. "Guidelines for evaluating and expressing the uncertainty of NIST measurement results", *NIST Technical Note 1297*, 1994 edition, U.S. Department of Commerce.

APPENDIX A: EVAPORATOR COILS AND CONDENSING UNIT DESCRIPTIONS

Appendix A presents specifications for the evaporators and condensing unit tested at NIST. It includes pictures, design data, and refrigerant circuitry representations in the input format of the EVAP-COND simulation package.

Matched System Coil

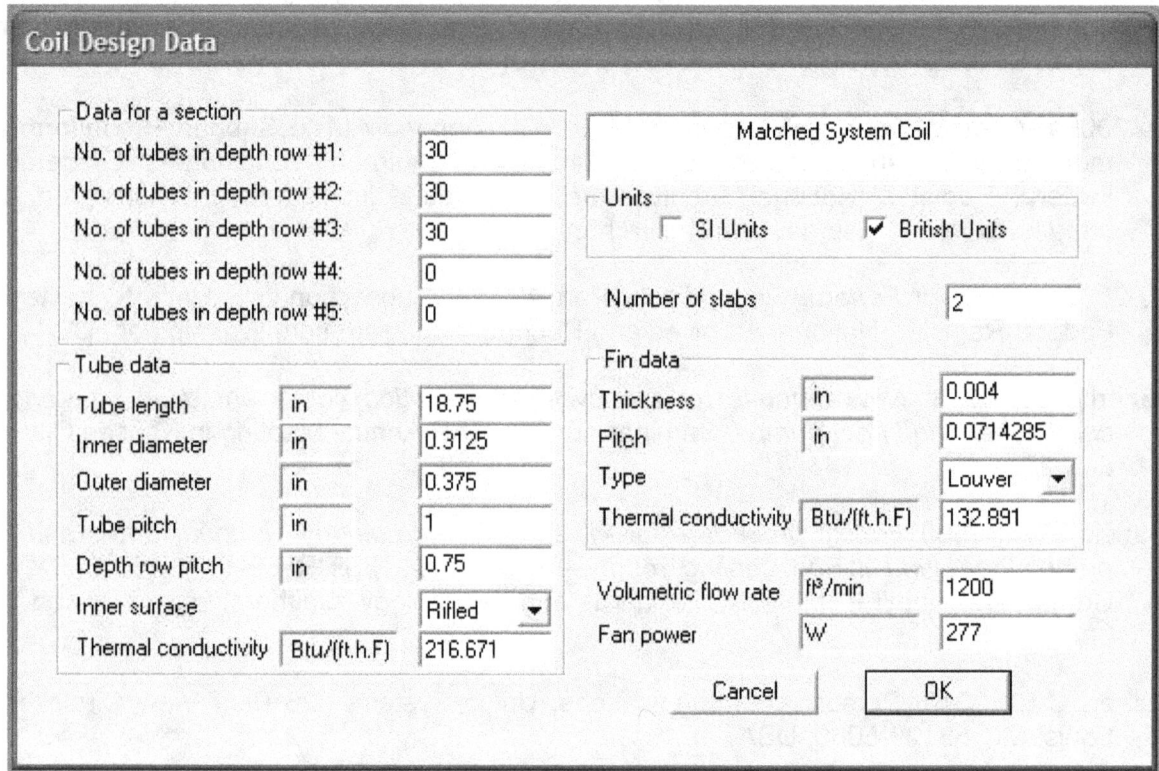

Figure A1: Matched coil description

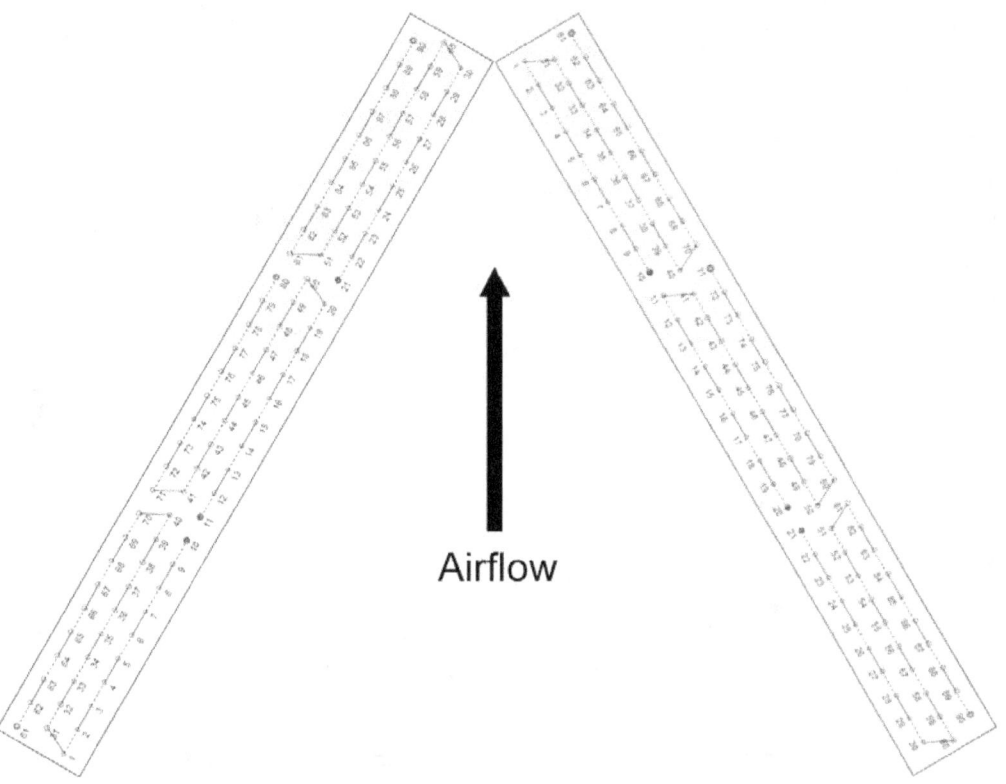

Airflow

Figure A2: Matched coil refrigerant circuitry

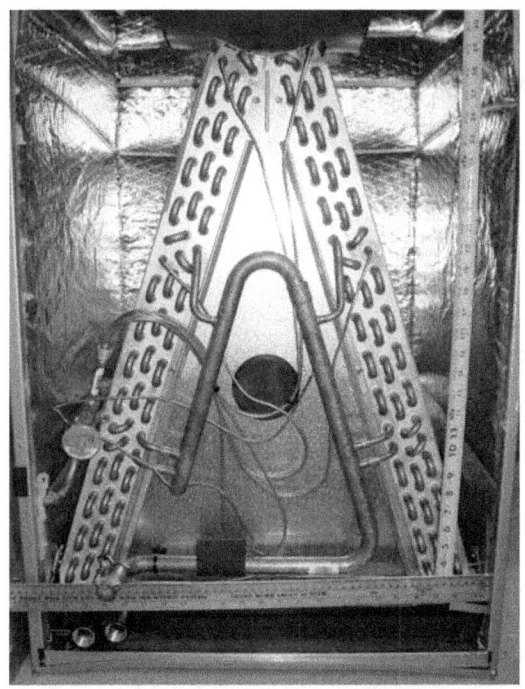

Figure A3: Matched coil side view

Mixed System Coil #1

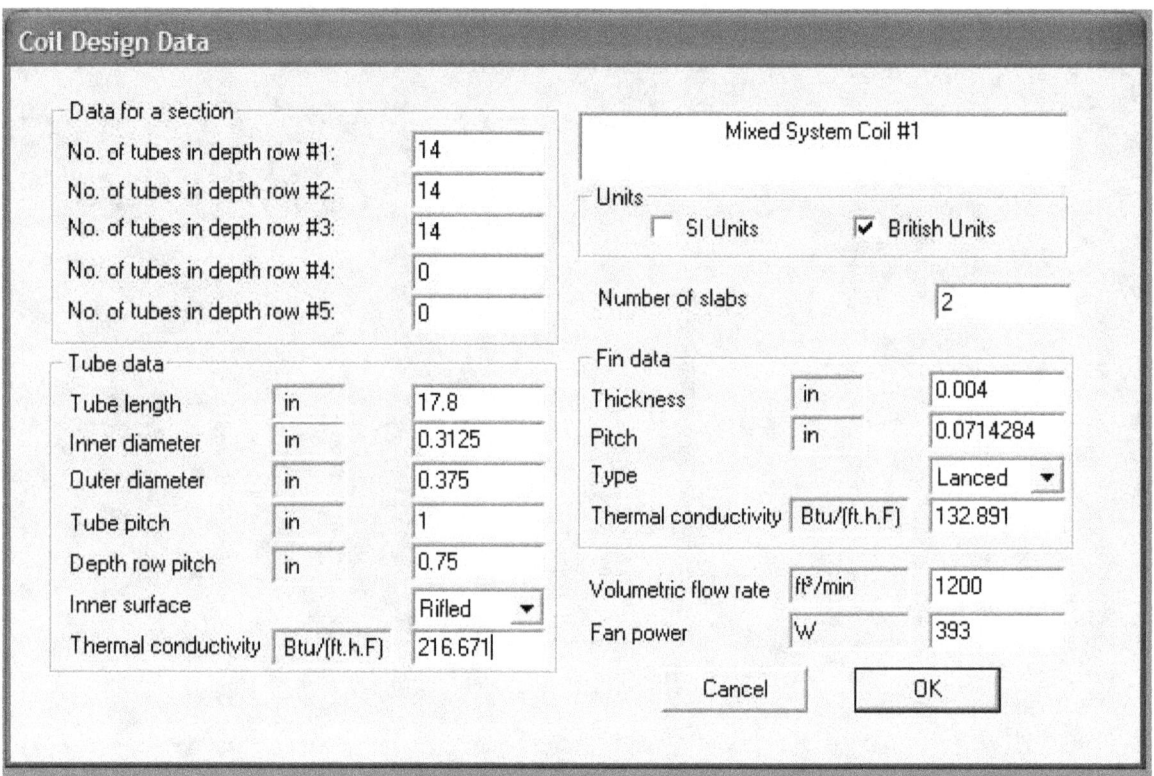

Figure A4: Mixed coil #1 description

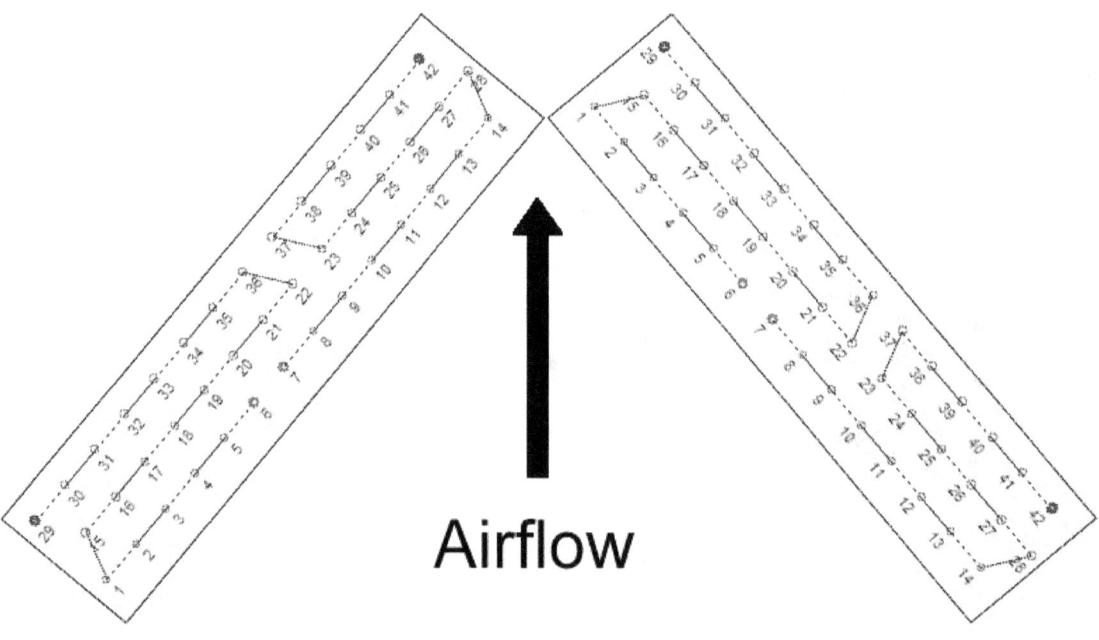

Airflow

Figure A5: Mixed coil #1 refrigerant circuitry

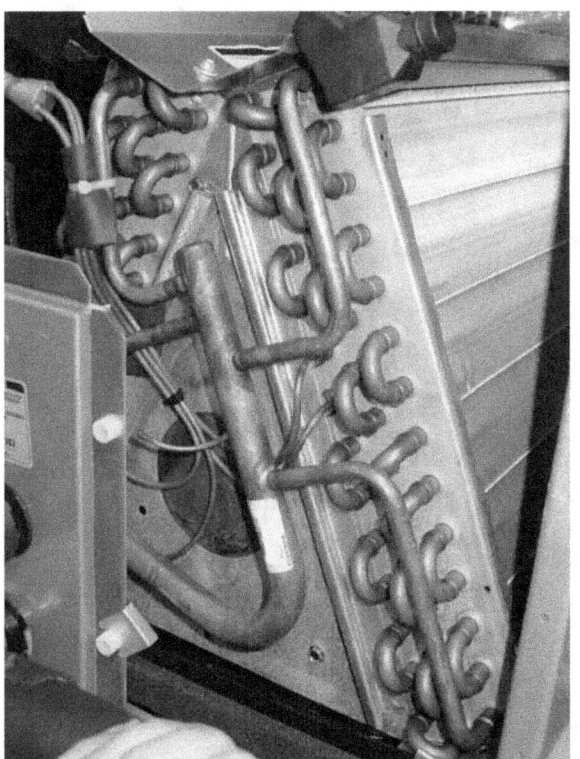

Figure A6: Mixed coil #1 side view

Mixed System Coil #2

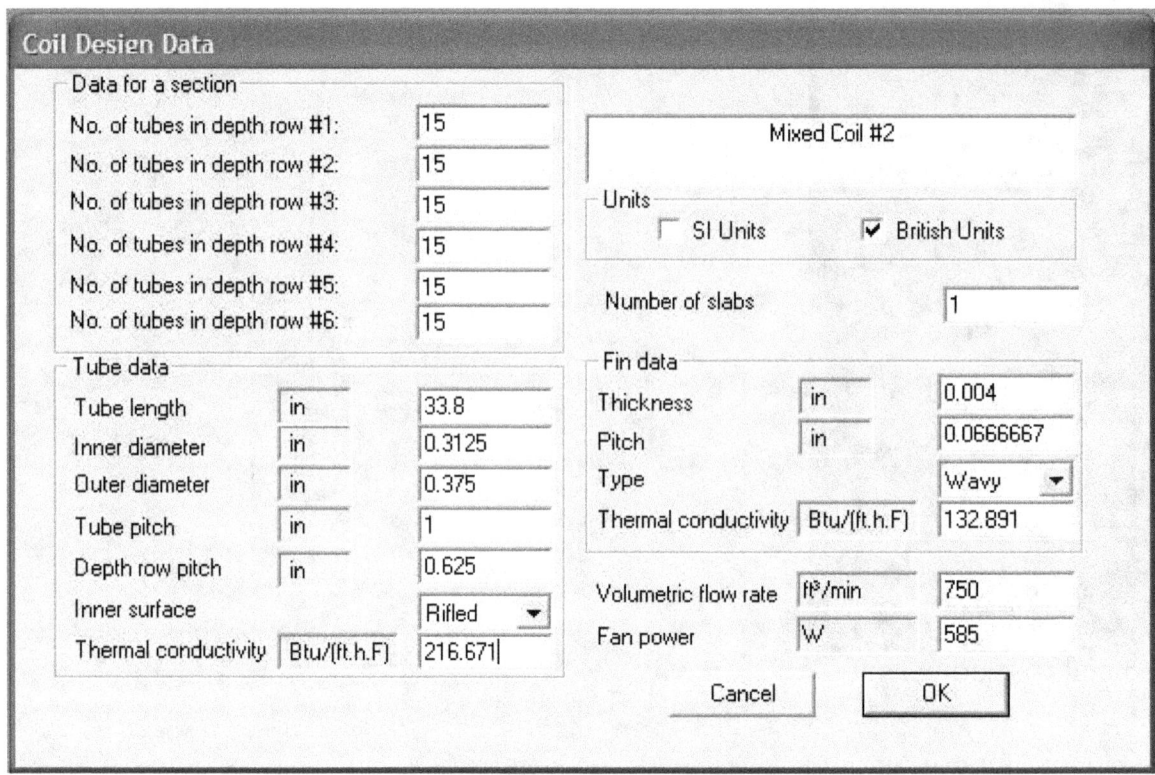

Figure A7: Mixed coil #2 description

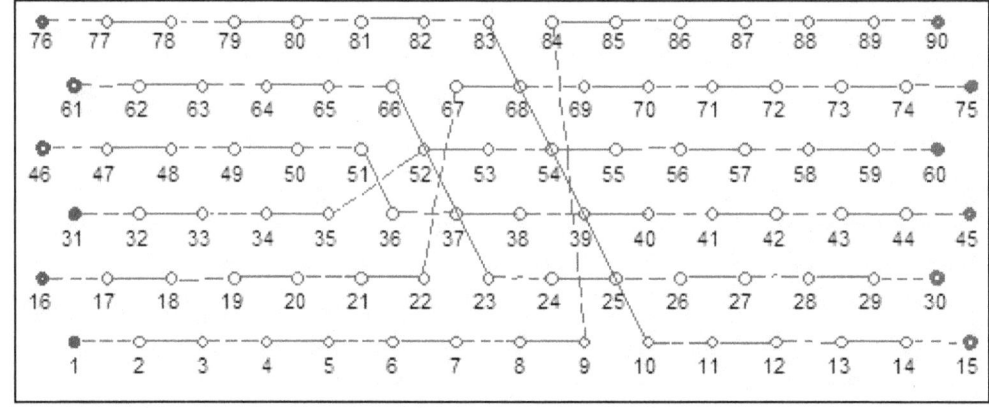

Airflow from the bottom

Figure A8: Mixed coil #2 refrigerant circuitry

Figure A9: Mixed coil #2 side views

Matched System Condenser

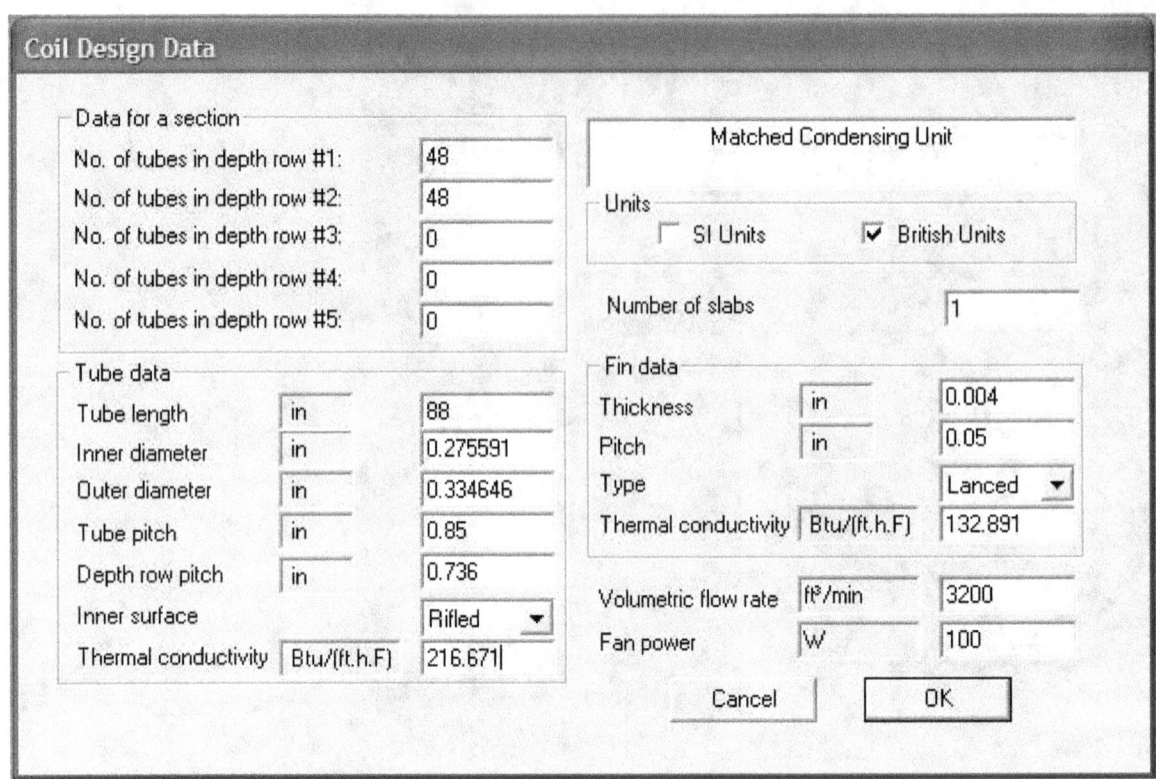

Figure A10: Matched condensing unit coil description

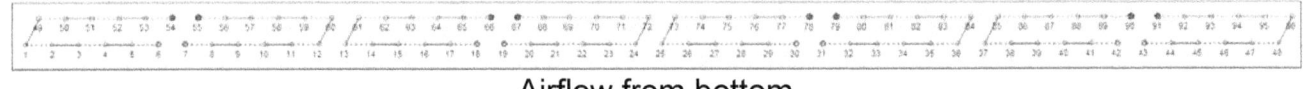

Airflow from bottom

Figure A11: Matched condensing unit coil refrigerant circuitry

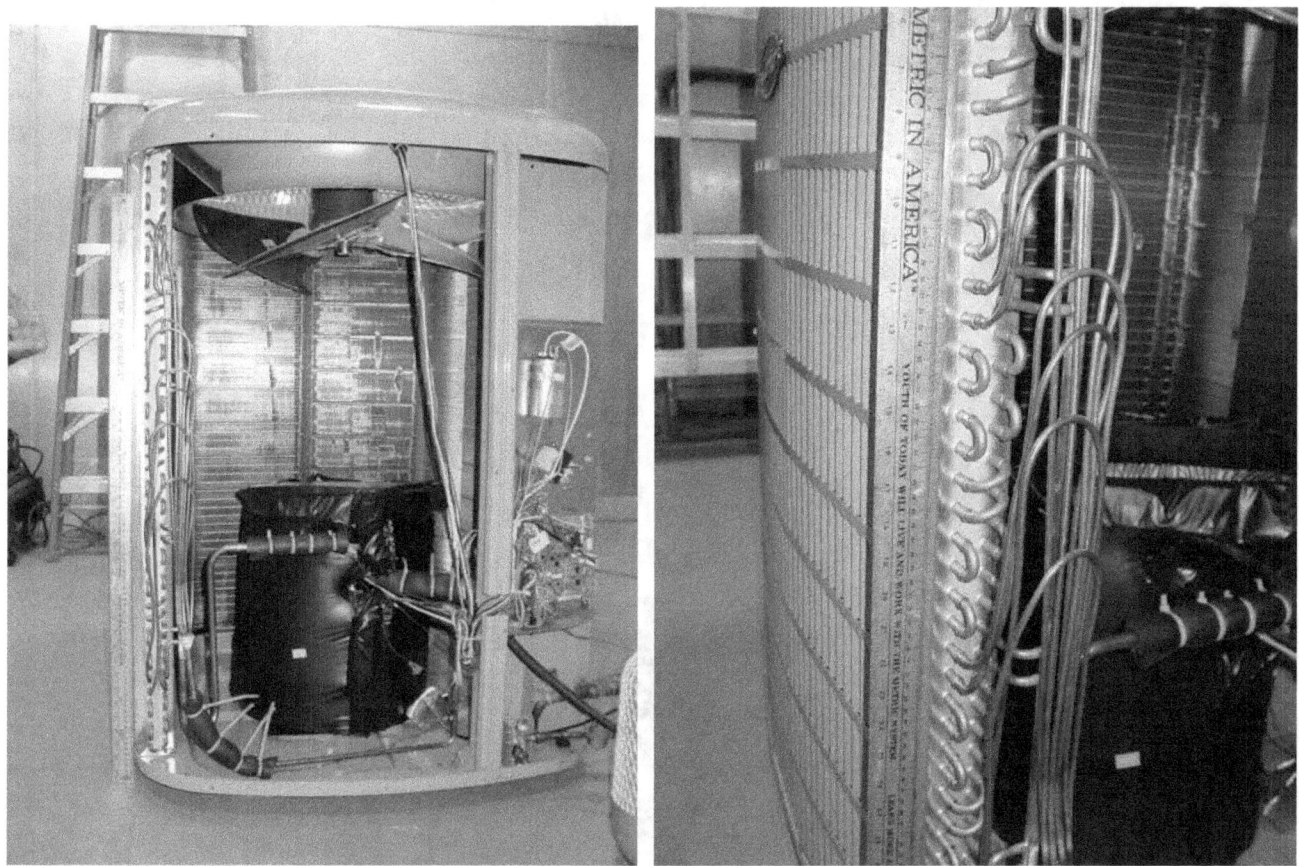
Figure A12: Matched condensing unit coil pictures

Figure B1: Water-cooled condensing unit with variable-speed, open-drive compressor

Figure B2: Oil separator arrangement

Figure B3: Rotameter/flowmeter used to adjust water flowrate to condenser heat exchanger

Figure B4: Rotameter/flowmeter used to adjust water flow to subcooler heat exchanger

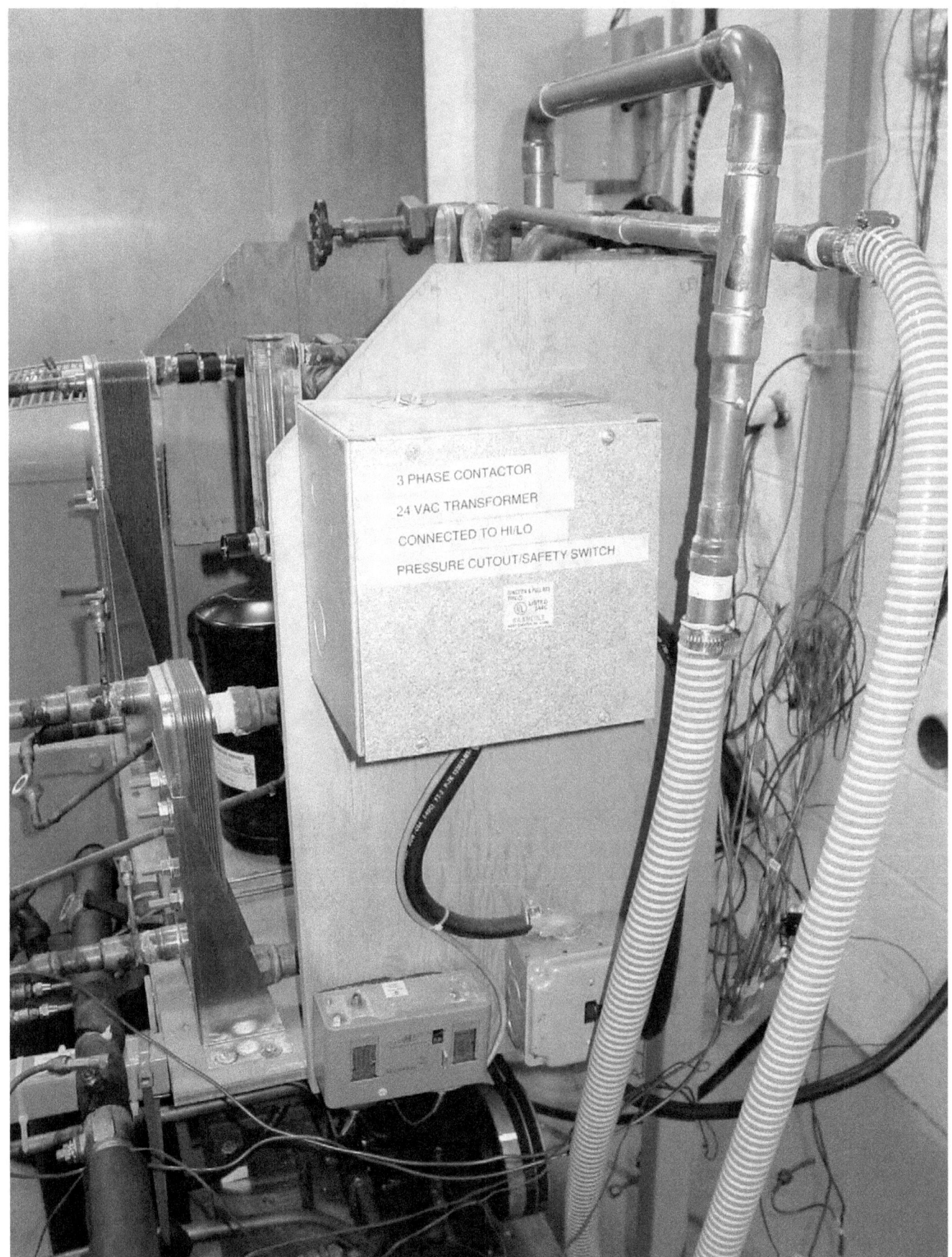

Figure B5: Right side view of water cooled condensing unit showing power contactor box, hi/lo pressure safety switch, and manual on/off switch

Figure B6: Brazed plate condenser heat exchanger

Figure B7: Brazed plate subcooler heat exchanger

72

Figure B8: Compressor and motor for water-cooled condensing unit (guard removed)

Figure B9: Open drive compressor name plate

73

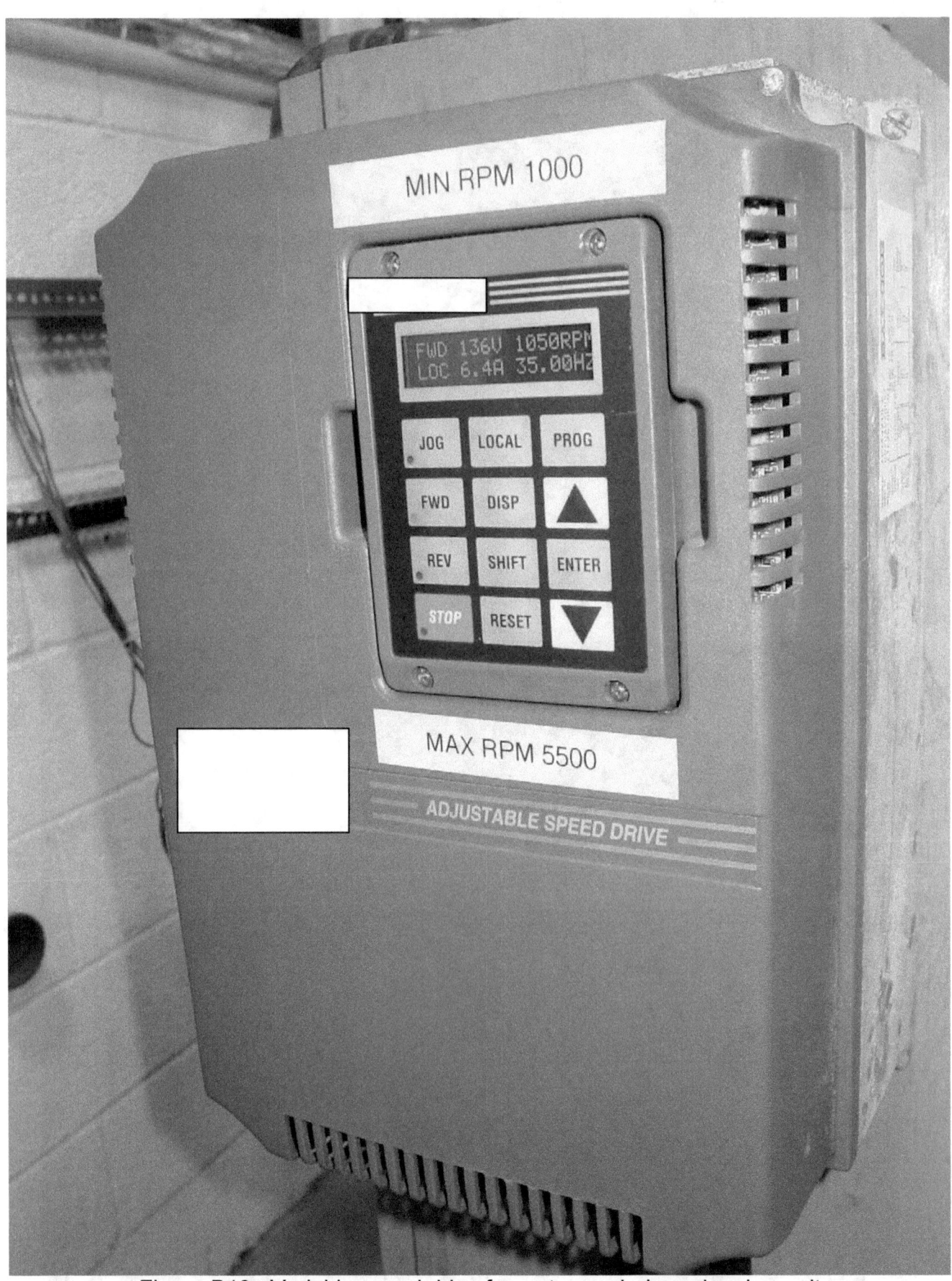

Figure B10: Variable speed drive for water-cooled condensing unit

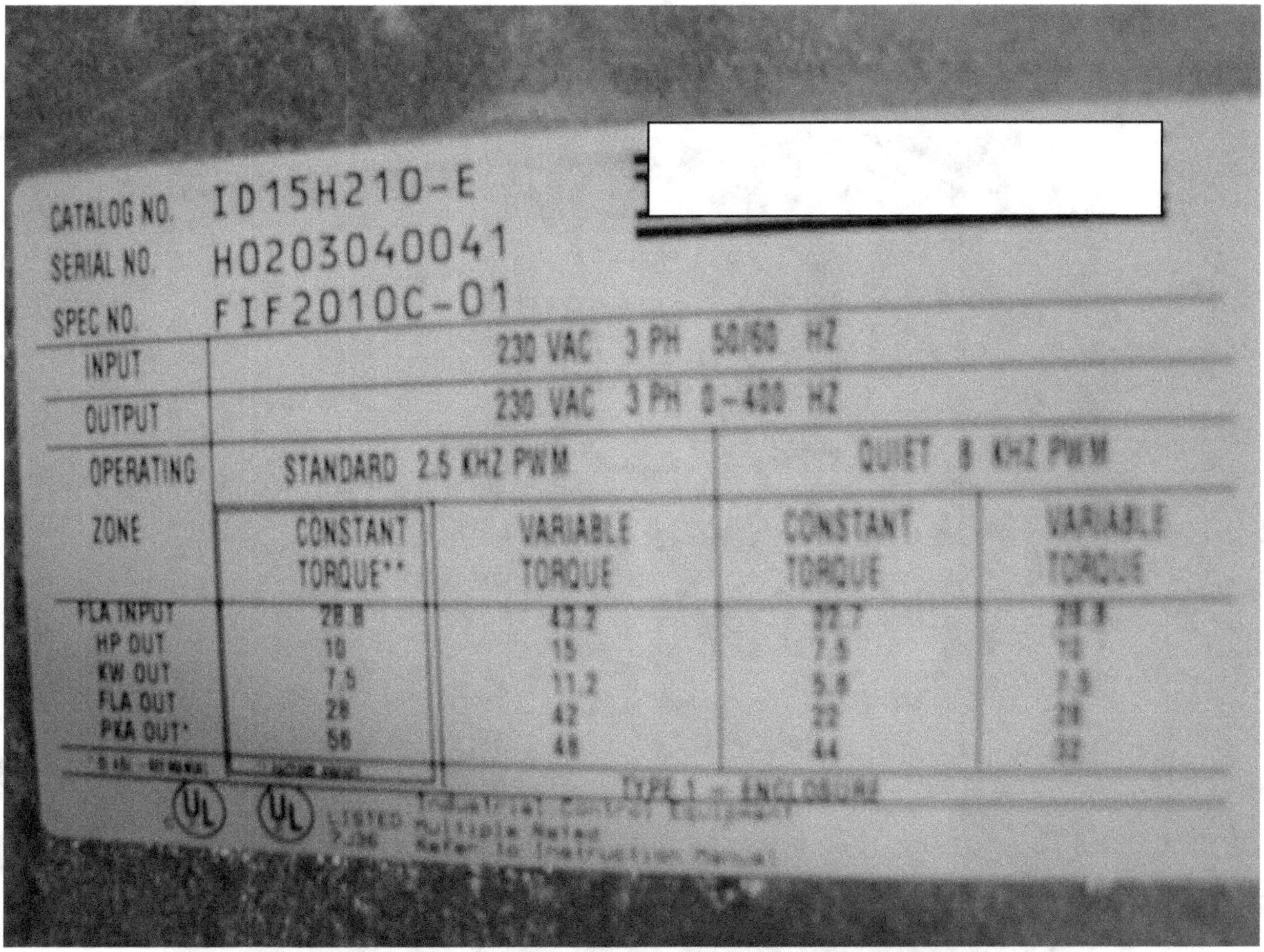

Figure B11: Variable speed drive name plate

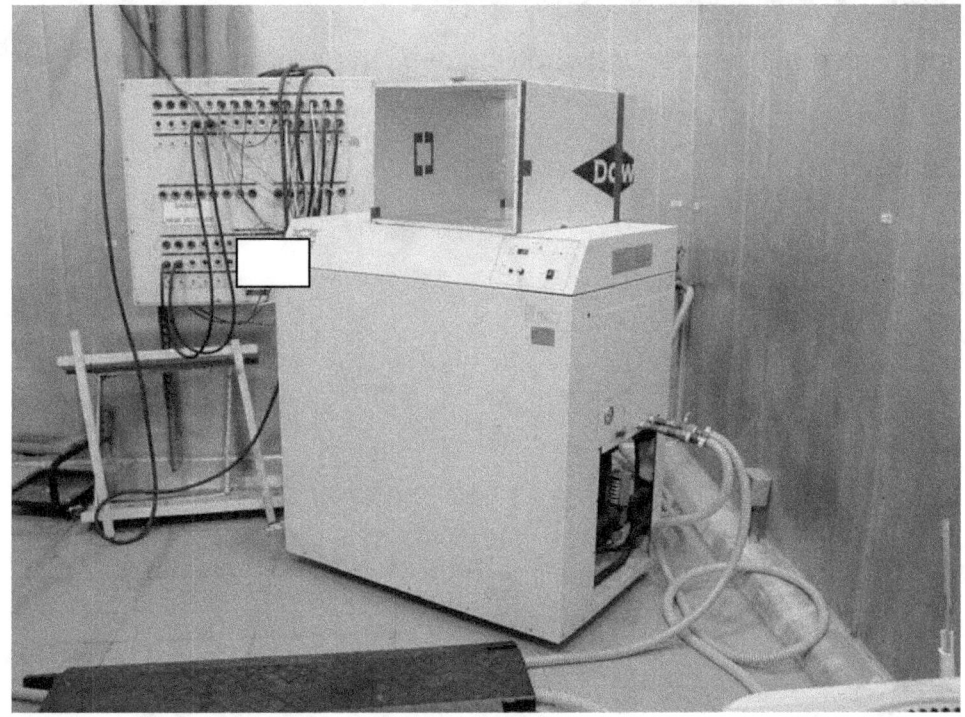

Figure B12: Portable water chiller connected to house water and used to control water temperature fed to the water-cooled condensing unit

Figure C1: Water-heated evaporator unit showing two evaporators and one superheater

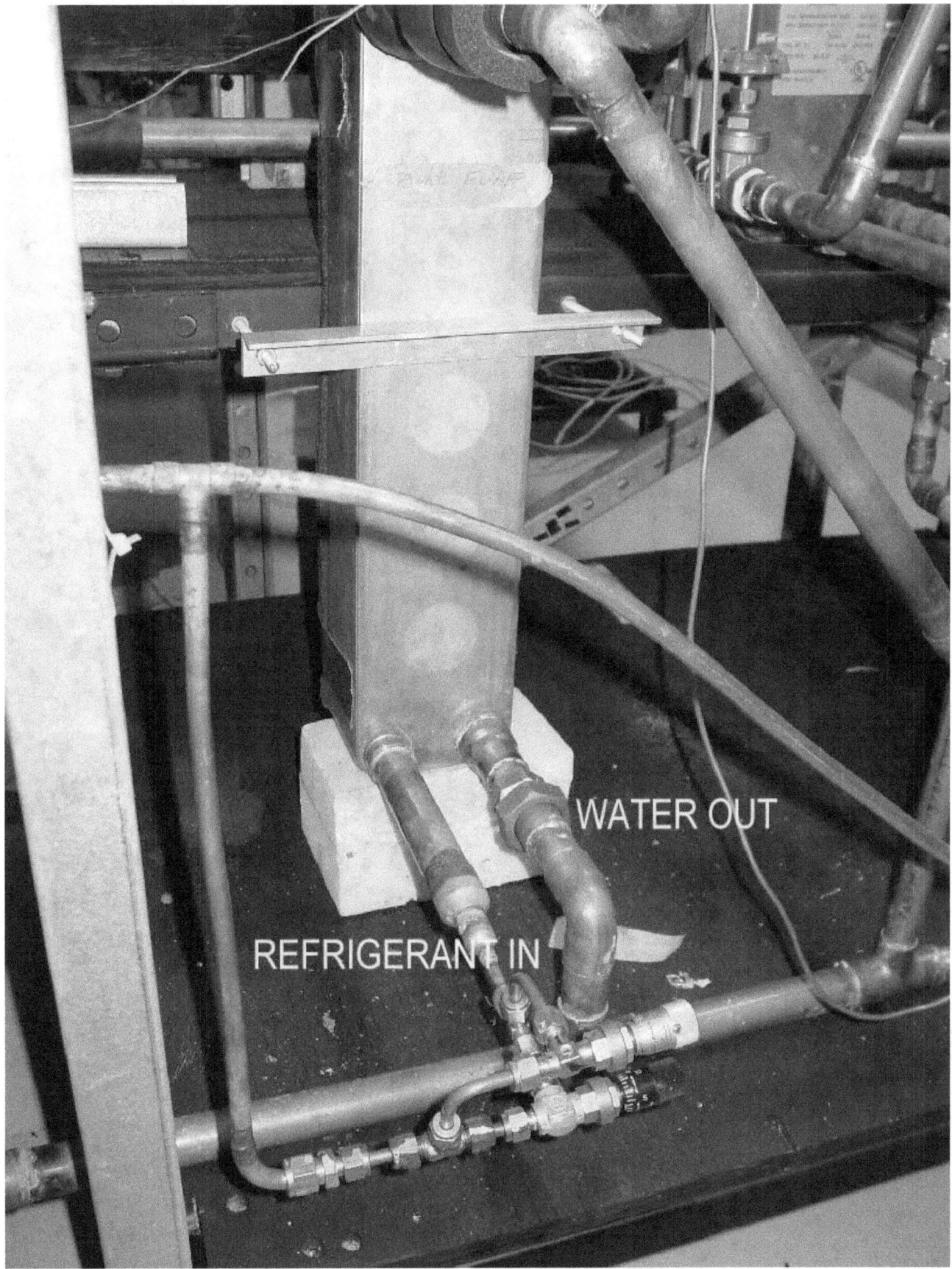

WATER OUT

REFRIGERANT IN

Figure C2: Left water-heated evaporator showing refrigerant expansion valves in parallel

Figure C3: Water-heated evaporator plate heat exchanger side-view

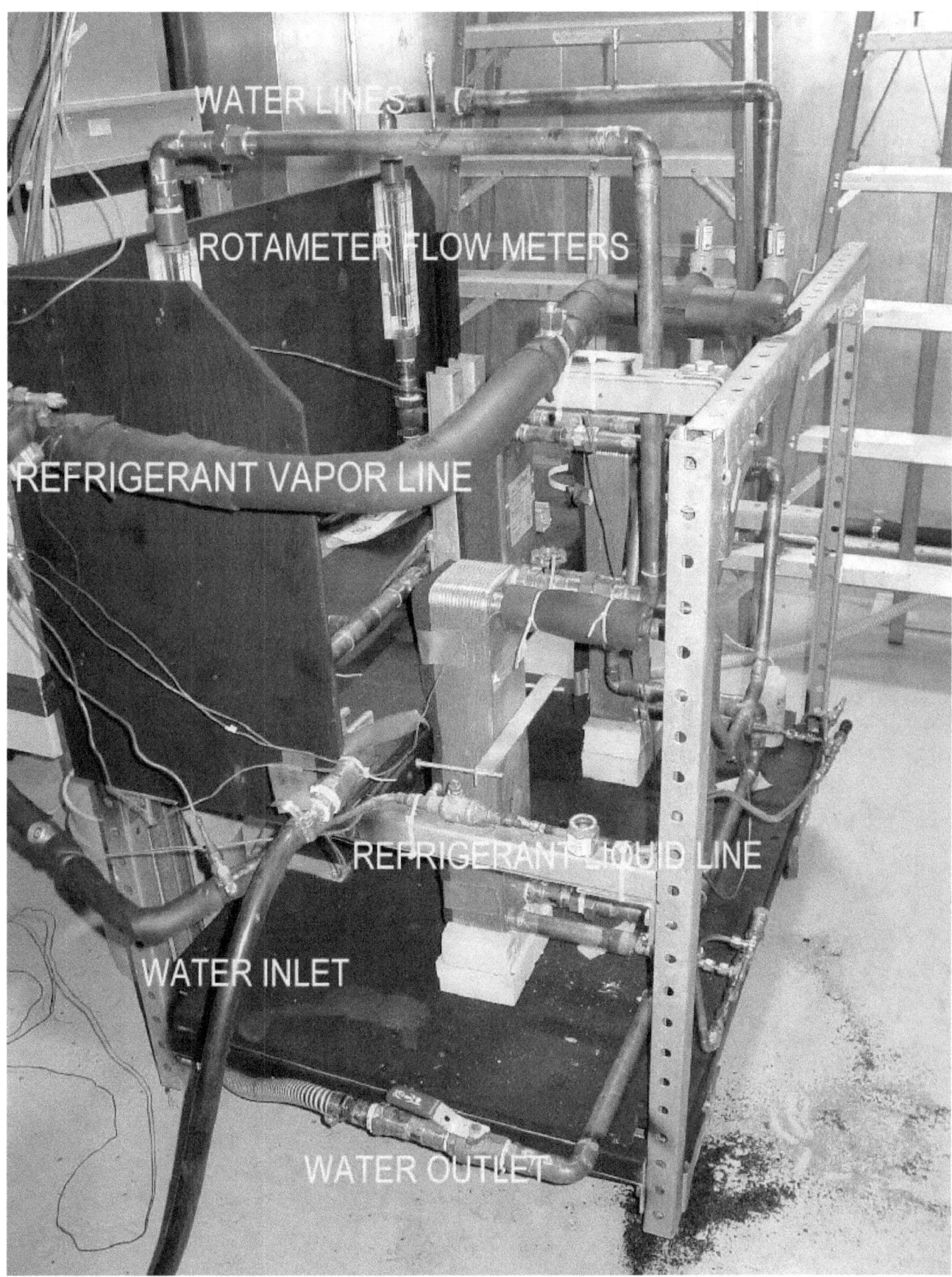

Figure C4: Water and refrigerant line connections

Figure C5: Superheater heat exchanger

Figure C6: Refrigerant expansion valve connnections

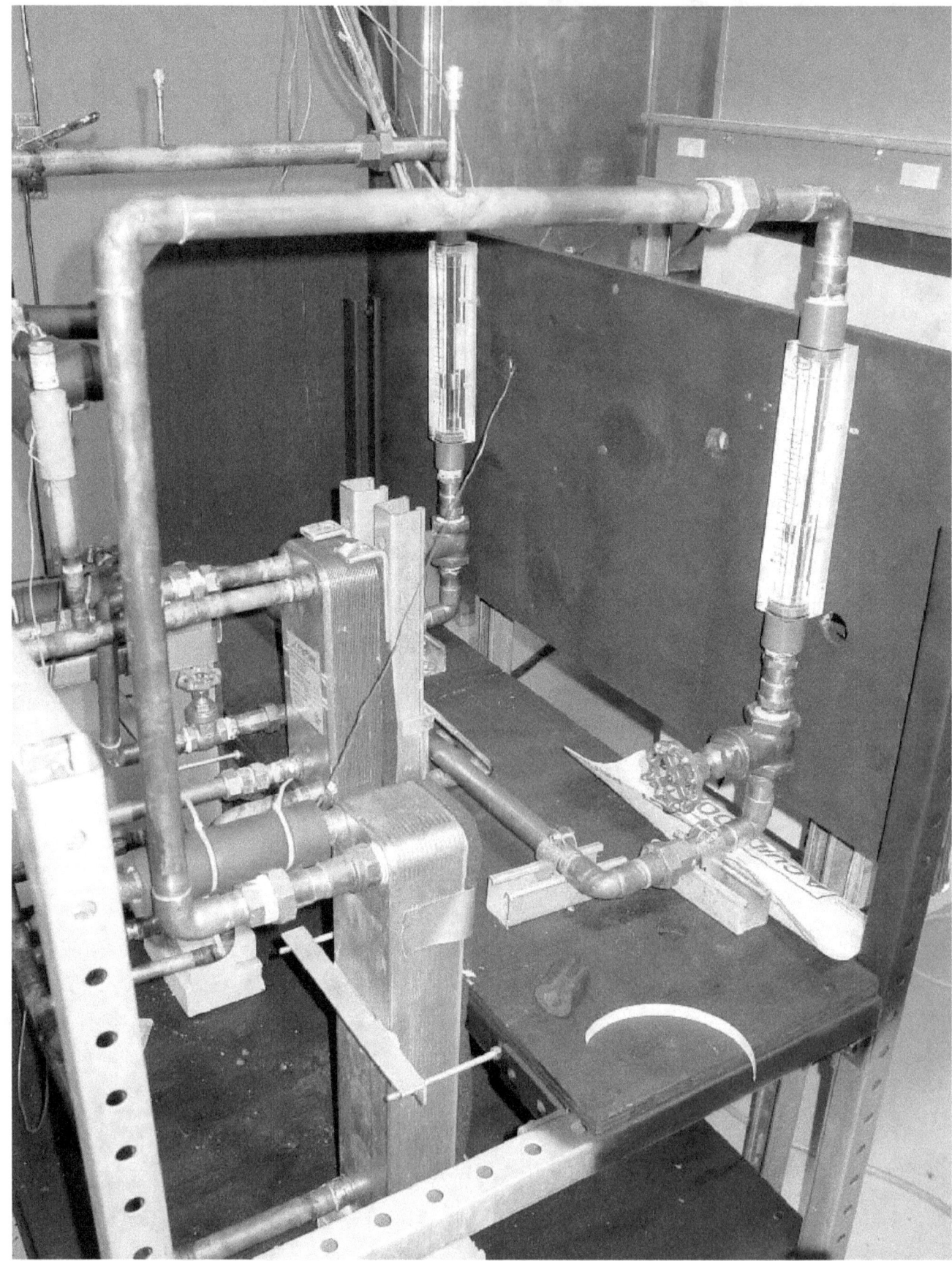

Figure C7: Right-side view of water-heated evaporator unit

Figure C8: Hot and cold house water mixed before going to water-heated evaporator unit

APPENDIX D: OBTAINING DATA USED IN THIS REPORT

Please contact Vance Payne for a copy of the data used to generate this report.

Vance Payne
National Institute of Standards and Technology
100 Bureau Drive, MS 8631
Gaithersburg, MD 20899

Email: vance.payne@nist.gov
Phone: 301-975-6663

www.ingramcontent.com/pod-product-compliance
Lightning Source LLC
Chambersburg PA
CBHW081828170526
45167CB00007B/2751